EnvStats

Steven P. Millard

EnvStats

An R Package for Environmental Statistics

 Springer

Steven P. Millard
Probability, Statistics and Information
Seattle, Washington
USA

ISBN 978-1-4614-8455-4   ISBN 978-1-4614-8456-1 (eBook)
DOI 10.1007/978-1-4614-8456-1
Springer New York Heidelberg Dordrecht London

Library of Congress Control Number: 2013945301

Printed on acid-free paper

Springer is part of Springer Science+Business Media (www.springer.com)

Preface

The environmental movement of the 1960s and 1970s resulted in the creation of several laws aimed at protecting the environment, and in the creation of federal, state, and local government agencies charged with enforcing these laws. Most of these laws mandate monitoring or assessment of the physical environment, which means someone has to collect, analyze, and explain environmental data. Numerous excellent books, guidance documents, and journal articles have been published to explain various aspects of applying statistical methods to these kinds of environmental data analyses. Also, several specialty software packages for specific niches in environmental statistics exist, such as for ecology, forestry, and climate modeling. Not very many software packages provide a comprehensive treatment of environmental statistics in the context of monitoring the physical environment as mandated by current environmental law.

EnvStats is an R package for environmental statistics. It is the open-source successor to the commercial module for S-Plus© called EnvironmentalStats for S-Plus, which was first released in April 1997. The EnvStats package, along with the R software environment, provides comprehensive and powerful software for environmental data analysis. EnvStats brings the major environmental statistical methods found in the literature and regulatory guidance documents into one statistical package, along with an extensive hypertext help system that explains *what* these methods do, *how* to use these methods, and *where* to find them in the environmental statistics literature. Also included are numerous built-in data sets from regulatory guidance documents and the environmental statistics literature. EnvStats combined with other R packages (e.g., for spatial analysis) provides the environmental scientist, statistician, researcher, and technician with tools to "get the job done!"

EnvStats and this user's manual are intended for anyone who has to make sense of environmental data, including statisticians, hydrologists, soil scientists, atmospheric scientists, geochemists, environmental engineers and consultants, hazardous and solid waste site managers, and regulatory agency analysts and enforcement officers. Some parts of EnvStats incorporate statistical methods that have appeared in the environmental literature but are not commonly found in any statistical software package. Some parts are specifically aimed at users who are required to collect and analyze environmental monitoring data in order to comply with federal and state Superfund, RCRA, CERCLA, and Subtitle D regulations for environmental monitoring at hazardous and solid waste sites. All of the functions in EnvStats, however, are useful to anyone who needs to analyze environmental data. In fact, all of these functions are useful to anyone who needs to analyze data.

This manual is divided into 9 chapters. Chapter 1 is an introduction to environmental statistics in general and the EnvStats package in particular, and includes information on system and user requirements, installing the software, loading and using the package, and getting technical support. The last section of the chapter includes a tutorial.

Chapters 2, 3, 4, 5, 6, 7, and 8 contain information about how to use the functions in EnvStats to design sampling programs and perform graphical and statistical analyses of environmental data. Chapter 9 shows you how to use EnvStats to perform Monte Carlo simulation and probabilistic risk assessment.

At the back of the book is an extensive list of references for environmental statistics as well as an index of key words and terms. In addition to using the index, you are encouraged to use the online hypertext help system as well.

Companion Scripts

Companion R scripts to reproduce the examples in this user's manual, as well as scripts for reproducing examples in US EPA guidance documents, are located in the **scripts** subdirectory of the directory where the package was installed. See Chap. 1 for more information.

Companion Textbook and Help Files

This user's manual provides *brief* explanations of various topics in environmental statistics. A companion textbook, currently in preparation and titled *Environmental Statistics with R* (Millard et al. 2014), provides more details and can be used as a textbook for a course in environmental statistics. (The predecessor to this textbook is Millard and Neerchal 2001.) Space constraints dictate that the examples in this user's manual convey a general sense of how various EnvStats functions can be used. The companion help files list all of the arguments associated with these functions and give more examples.

Technical Support

Technical support for R is available through the R-help mailing list (see the URL www.r-project.org for more information). Technical support for questions or problems specifically related to the functioning of the EnvStats package is available by e-mailing the author at EnvStats@ProbStatInfo.com.

Typographic Conventions

Throughout this user's manual, the following typographic conventions are used:

- The **bold font** is used for chapter and section headings, as well as operating system commands and file names. Sometimes it is also used for emphasis. R menu selections are shown in an abbreviated form using this font and the arrow symbol (>) to indicate a selection within a menu, as in **Packages>Load package…**.
- The *italic courier font* is used to display what you type within an R Command or Script Window.
- The `courier font` is used to display output from R and the names of R objects.
- The *italic font* is used for chapter and help file titles within the text, emphasis, and user-supplied variables within R commands.
- The ***bold italic font*** is used for emphasis.

R commands are preceded with the "greater than" sign, i.e., >, which is the default R prompt. For commands that require more than one line of input, the line or lines following the first line are indented, whereas within R they are preceded with the "plus" sign, i.e., +, which is the default R continuation prompt. Note that page size and formatting for this book determine how command lines are split. As a user, you can choose to split lines differently or not at all.

A Note About Figure Titles

To conserve space, very few of the plots shown in this user's manual have titles above them since all plots are labeled at the bottom with a figure number and title. However, several plotting functions in **EnvStats** produce figures with titles at the top by default if the argument `main` is not supplied. Thus, there are several examples in this manual where the figure shown does not include a title at the top, but if you type in the commands as shown or run the companion script to produce the figure, you will produce a figure with a title at the top. To create a different title from the one shown, use the `main` argument or, to suppress the title, set the argument `main=""`.

Acknowledgments

In the early 1980s, while pursuing a degree in biostatistics, I became aware of a knowledge gap in the field of environmental statistics. There was lots of research going on in the academic field of environmental statistics, but there were lots of poor designs and analyses being carried out in the real-world "field." There were even federal laws mandating incorrect statistical analyses of groundwater samples at hazardous waste sites.

One of the first steps to improving the quality of environmental statistics is to improve the quality of communication between statisticians and experts in various environmental fields. Several researchers and practitioners have done this by publishing excellent books and journal articles dealing with general and specific problems in environmental statistics (see the references listed at the back of this book).

The next logical step is to provide the necessary tools to carry out all the great methods and ideas in the literature. In the mid-1980s, I sat down with two friends of mine, Dennis Lettenmaier and Jim Hughes, to talk about building a software package for environmental statistics. Dennis and Jim gave me some good ideas, but I never acted on these ideas because at the time I didn't feel like I had the tools to make the kind of software package I wanted to.

The emergence of S-Plus© and help-authoring tools in the early 1990s changed all that. S-Plus at that time was one of the few statistical software packages with great graphics that allowed users to write their own functions and create pull-down menus. ForeHelp© was one of several help-authoring tools that allowed anyone who can figure out a word processor to write a hypertext Windows, HTML, or JAVA help system. The availability of these tools resulted in the creation of EnvironmentalStats for S-Plus, which was first released in April 1997. The emergence of open-source software, including R, has led to a revolution in the availability of specialty tools for data analysis. Porting EnvironmentalStats for S-Plus to the R environment in the form of the EnvStats package, where it can be integrated with hundreds of other R packages, is the next logical step in its evolution.

There are several people who helped and encouraged me over the past several years as I developed and extended first EnvironmentalStats for S-Plus and then the R package EnvStats. Most recently, thanks to Charles Davis (Envirostat & Statistics Ltd) and Kirk Cameron (MacStat Consulting Ltd) for help with updating the functions for simultaneous nonparametric prediction intervals, and to Phil Dixon (Iowa State University) for early feedback on EnvStats.

Thanks to Insightful Corporation (now part of TIBCO) for their past marketing, distribution, and support of EnvironmentalStats for S-Plus. Several people at Insightful contributed to the success of EnvironmentalStats for S-Plus, including Kim Leader, Scott Blachowicz, Christopher Disdero, Cheryl Mauer, Patrick Aboyoun, Tim Wegner, Rich Calaway, Charlie Roosen, Stephen Kaluzny, and the Insightful technical support team. Special thanks to the members of the S-news group for constantly giving me insight into the workings of S-Plus. I would also like to thank Gilbert FitzGerald, former Director of Product Development at MathSoft, for his encouragement and advice. Thanks to Jim Hughes (University of Washington) for being a statistical consultant's statistical consultant. Also thanks to Tim Cohn and Dennis Helsel (US Geological Survey), Chris Fraley (Insightful and University of Washington), Jon Hosking (IBM Research Division),

Ross Prentice (Fred Hutchinson Cancer Research Center), and Terry Therneau (Mayo Clinic) for their help in answering technical questions. Dick Gilbert was a constant source of encouragement and advice – thanks! Thanks to the Beta testers of EnvironmentalStats for S-Plus for their feedback on early versions.

There would be no EnvStats package for R unless there was R, so I am indebted to Ross Ihaka and Robert Gentleman, the creators of R, and the R development core team. There would have been no EnvironmentalStats for S-Plus if there had been no S-Plus, so I am grateful to Doug Martin for leading the creation of S-Plus. There would be no S-Plus and no R if there had not first been S. I am grateful to the researchers of the Bell Laboratories S team at AT&T who first created S, including Richard A. Becker, John M. Chambers, Alan R. Wilks, William S. Cleveland, and others.

I am grateful to Marc Strauss and Hannah Bracken at Springer-Verlag for their help in transforming this user's manual into a book. Finally, and most gratefully, thanks to my wife Stacy and my son Chris for their moral support. I couldn't have done it without you!

EnvStats is the culmination of a dream I had over 25 years ago. I hope it provides you with the tools you need to "get the job done," and I hope you enjoy it as much as I have enjoyed creating it!

Seattle, WA
March 2013

Steven P. Millard

Contents

Preface v
    Companion Scripts vi
    Companion Textbook and Help Files vi
    Technical Support vi
    Typographic Conventions vii
    A Note About Figure Titles vii
    Acknowledgments vii

1   Getting Started 1
    1.1   Introduction 1
    1.2   What Is Environmental Statistics? 1
    1.3   What Is EnvStats? 2
    1.4   Intended Audience and Users 3
    1.5   System Requirements 4
    1.6   Installing EnvStats 4
    1.7   Starting EnvStats 4
    1.8   Getting Help and Using Companion Scripts 5
    1.9   A Note About Examples and Masking 6
    1.10 Unloading EnvStats 7
    1.11 A Tutorial 7
         1.11.1   The TcCB Data 8
         1.11.2   Computing Summary Statistics 9
         1.11.3   Looking at the TcCB Data 10
         1.11.4   Quantile (Empirical CDF) Plots 12
         1.11.5   Assessing Goodness-of-Fit with Quantile-Quantile
                  Plots 14
         1.11.6   Estimating Distribution Parameters 16
         1.11.7   Testing for Goodness of Fit 19
         1.11.8   Estimating Quantiles and Computing Confidence
                  Limits 22
         1.11.9   Comparing Two Distributions Using Nonparametric
                  Tests 23
    1.12 Summary 24

2 Designing a Sampling Program **25**
  2.1   Introduction 25
  2.2   The Necessity of a Good Sampling Design 25
  2.3   What Is a Population and What Is a Sample? 25
  2.4   Random Versus Judgment Sampling 26
  2.5   Common Mistakes in Environmental Studies 26
  2.6   The Data Quality Objectives Process 27
  2.7   Power and Sample Size Calculations 28
  2.8   Sample Size for Confidence Intervals 28
      2.8.1   Confidence Interval for the Mean of a
                  Normal Distribution 30
      2.8.2   Confidence Interval for a Binomial Proportion 34
      2.8.3   Nonparametric Confidence Interval for a Percentile 38
  2.9   Sample Size for Prediction Intervals 40
      2.9.1   Prediction Interval for a Normal Distribution 40
      2.9.2   Nonparametric Prediction Interval 43
  2.10 Sample Size for Tolerance Intervals 44
      2.10.1   Tolerance Interval for a Normal Distribution 45
      2.10.2   Nonparametric Tolerance Interval 47
  2.11 Sample Size and Power for Hypothesis Tests 49
      2.11.1   Testing the Mean of a Normal Distribution 51
      2.11.2   Testing a Binomial Proportion 55
      2.11.3   Testing Multiple Wells for Compliance with
                  Simultaneous Prediction Intervals 57
  2.12 Summary 61

3 Looking at Data **63**
  3.1   Introduction 63
  3.2   EDA Using ENVSTATS 64
  3.3   Summary Statistics 65
      3.3.1   Summary Statistics for TcCB Concentrations 65
  3.4   Strip Charts 66
  3.5   Empirical PDF Plots 66
  3.6   Quantile (Empirical CDF) Plots 67
      3.6.1   Empirical CDFs for the TcCB Data 68
  3.7   Probability Plots or Quantile-Quantile (Q-Q) Plots 68
      3.7.1   Q-Q Plots for the Normal and Lognormal
                  Distribution 69
      3.7.2   Q-Q Plots for Other Distributions 70
      3.7.3   Using Q-Q Plots to Compare Two Data Sets 72
      3.7.4   Building an Internal Gestalt for Q-Q Plots 73
  3.8   Box-Cox Data Transformations and Q-Q Plots 75
  3.9   Summary 78

4   Probability Distributions..**79**
   4.1   Introduction...79
   4.2   Probability Density Function (PDF)...88
         4.2.1   Probability Density Function for Lognormal
                 Distribution..88
         4.2.2   Probability Density Function for a Gamma
                 Distribution..90
   4.3   Cumulative Distribution Function (CDF).....................................92
         4.3.1   Cumulative Distribution Function for Lognormal
                 Distribution..92
   4.4   Quantiles and Percentiles..93
         4.4.1   Quantiles for Lognormal Distribution..............................93
   4.5   Generating Random Numbers 94
         4.5.1   Generating Random Numbers from a Univariate
                 Distribution..94
         4.5.2   Generating Multivariate Normal Random
                 Numbers...94
         4.5.3   Generating Multivariate Observations Based
                 on Rank Correlations..95
   4.6   Summary..96

5   Estimating Distribution Parameters and Quantiles **97**
   5.1   Introduction...97
   5.2   Estimating Distribution Parameters..97
         5.2.1   Estimating Parameters of a Normal Distribution 97
         5.2.2   Estimating Parameters of a Lognormal
                 Distribution..99
         5.2.3   Estimating Parameters of a Gamma Distribution.............101
         5.2.4   Estimating the Parameter of a Binomial
                 Distribution..102
   5.3   Estimating Distribution Quantiles 102
         5.3.1   Estimating Quantiles of a Normal Distribution................103
         5.3.2   Estimating Quantiles of a Lognormal Distribution...........105
         5.3.3   Estimating Quantiles of a Gamma Distribution 106
         5.3.4   Nonparametric Estimates of Quantiles............................107
   5.4   Summary..112

6   Prediction and Tolerance Intervals..**113**
   6.1   Introduction...113
   6.2   Prediction Intervals...113
         6.2.1   Prediction Intervals for a Normal Distribution................116
         6.2.2   Prediction Intervals for a Lognormal Distribution...........119
         6.2.3   Prediction Intervals for a Gamma Distribution 122
         6.2.4   Nonparametric Prediction Intervals................................125

6.3   Simultaneous Prediction Intervals 128
    6.3.1   Simultaneous Prediction Intervals for a Normal
           Distribution 131
    6.3.2   Simultaneous Prediction Intervals for a Lognormal
           Distribution 133
    6.3.3   Simultaneous Prediction Intervals for a Gamma
           Distribution 136
    6.3.4   Simultaneous Nonparametric Prediction Intervals 136
6.4   Tolerance Intervals 141
    6.4.1   Tolerance Intervals for a Normal Distribution 142
    6.4.2   Tolerance Intervals for a Lognormal Distribution.............144
    6.4.3   Tolerance Intervals for a Gamma Distribution.................146
    6.4.4   Nonparametric Tolerance Intervals 146
6.5   Summary..148

7   **Hypothesis Tests**...**149**
7.1   Introduction..149
7.2   Goodness-of-Fit Tests..149
    7.2.1   One-Sample Goodness-of-Fit Tests for
           Normality 151
    7.2.2   Testing Several Groups for Normality 154
    7.2.3   One-Sample Goodness-of-Fit Tests for Other
           Distributions 159
    7.2.4   Two-Sample Goodness-of-Fit Test to Compare
           Samples...161
7.3   One-, Two-, and k-Sample Comparison Tests................................163
    7.3.1   Two- and k-Sample Comparisons for Location.................164
    7.3.2   Chen's Modified One-Sample t-Test for Skewed
           Data..166
    7.3.3   Two-Sample Linear Rank Tests and the Quantile
           Test 168
7.4   Testing for Serial Correlation 169
7.5   Testing for Trend 169
    7.5.1   Testing for Trend in the Presence of Seasons 170
7.6   Summary..172

8   **Censored Data** **175**
8.1   Introduction..175
8.2   Classification of Censored Data 175
8.3   Functions for Censored Data 176
8.4   Graphical Assessment of Censored Data.......................................176
    8.4.1   Quantile (Empirical CDF) Plots for Censored Data..........176
    8.4.2   Comparing an Empirical and Hypothesized CDF 179
    8.4.3   Comparing Two Empirical CDFs.................................180

	8.4.4	Q-Q Plots for Censored Data	181
	8.4.5	Box-Cox Transformations for Censored Data	183
8.5	Estimating Distribution Parameters		184
	8.5.1	The Normal and Lognormal Distribution	184
	8.5.2	The Lognormal Distribution (Original Scale)	188
	8.5.3	The Gamma Distribution	190
	8.5.4	Estimating the Mean Nonparametrically	191
8.6	Estimating Distribution Quantiles		193
	8.6.1	Parametric Estimates of Quantiles	194
	8.6.2	Nonparametric Estimates of Quantiles	198
8.7	Prediction Intervals		198
	8.7.1	Parametric Prediction Intervals	198
	8.7.2	Nonparametric Prediction Intervals	200
8.8	Tolerance Intervals		200
	8.8.1	Parametric Tolerance Intervals	200
	8.8.2	Nonparametric Tolerance Intervals	203
8.9	Hypothesis Tests		203
	8.9.1	Goodness-of-Fit Tests	203
	8.9.2	Nonparametric Tests to Compare Two Groups	207
8.10	Summary		209
9	**Monte Carlo Simulation and Risk Assessment**		**211**
9.1	Introduction		211
9.2	Overview		212
9.3	Monte Carlo Simulation		212
	9.3.1	Simulating the Distribution of the Sum of Two Normal Random Variables	213
9.4	Generating Random Numbers		216
	9.4.1	Generating Random Numbers from a Uniform Distribution	216
	9.4.2	Generating Random Numbers from an Arbitrary Distribution	216
	9.4.3	Latin Hypercube Sampling	217
	9.4.4	Example of Simple Random Sampling versus Latin Hypercube Sampling	220
	9.4.5	Properties of Latin Hypercube Sampling	222
	9.4.6	Generating Correlated Multivariate Random Numbers	222
9.5	Uncertainty and Sensitivity Analysis		224
	9.5.1	Important Versus Sensitive Parameters	225
	9.5.2	Uncertainty Versus Variability	226
	9.5.3	Sensitivity Analysis Methods	227
	9.5.4	Uncertainty Analysis Methods	230
	9.5.5	Caveat	232

9.6   Risk Assessment 232
       9.6.1   Definitions 232
       9.6.2   Building a Risk Assessment Model....................................235
       9.6.3   Example: Quantifying Variability and Parameter
               Uncertainty 236
9.7   Summary..241

References..**243**

Index...**283**

Chapter 1

Getting Started

1.1 Introduction

Welcome to EnvStats! This user's manual provides step-by-step guidance to using this software. EnvStats is an R package for environmental statistics. This chapter is an introduction to environmental statistics in general and EnvStats in particular, and includes information on system and user requirements, installing the software, loading and using the package, and getting technical support. The last section of the chapter provides a brief tutorial.

1.2 What Is Environmental Statistics?

Environmental statistics is the application of statistical methods to problems concerning the environment. Examples of activities that require the use of environmental statistics include:

- Monitoring air or water quality.
- Monitoring groundwater quality near a hazardous or solid waste site.
- Using risk assessment to determine whether a potentially contaminated area needs to be cleaned up, and, if so, how much.
- Assessing whether a previously contaminated area has been cleaned up according to some specified criterion.
- Using hydrological data to predict the occurrences of floods.

The term "environmental statistics" also includes work done in atmospheric and climate-change research and modeling, and the application of statistics in the fields of ecology, geology, chemistry, epidemiology, and oceanography. This user's manual concentrates on statistical methods to analyze chemical concentrations and physical parameters, usually in the context of mandated environmental monitoring.

Environmental statistics is a special field of statistics. Probability and statistics deal with situations in which the outcome is not certain. They are built upon the concepts of a ***population*** and a ***sample*** from the population. ***Probability*** deals with predicting the characteristics of the sample, given that you know the characteristics of the population (e.g., the probability of picking an ace out of a deck of 52 well-shuffled standard playing cards). ***Statistics*** deals with inferring the characteristics of the population, given information from one or more samples from the population (e.g., estimating the concentration of dissolved oxygen in a lake based on samples taken at various depths and locations).

S.P. Millard, *EnvStats: An R Package for Environmental Statistics*,
DOI 10.1007/978-1-4614-8456-1_1, © Springer Science+Business Media New York 2013

The field of environmental statistics is relatively young and employs several statistical methods that have been developed in other fields of statistics, such as sampling design, exploratory data analysis, basic estimation and hypothesis testing, quality control, multiple comparisons, survival analysis, and Monte Carlo simulation. Nonetheless, special problems have motivated innovative research, and both traditional and new journals now report on statistical methods that have been developed in the context of environmental monitoring (see the references listed at the end of this book.)

In addition, environmental legislation such as the Clean Air Act, the Clean Water Act, the National Environmental Policy Act (NEPA), the Occupational Safety and Health Act, the Federal Insecticide, Fungicide, and Rodenticide Act (FIFRA), the Comprehensive Emergency Response, Compensation, and Liability Act (CERCLA), the Resource and Recovery Act (RCRA) and all of their subsequent amendments have spawned environmental regulations and agency guidance documents that mandate or suggest various statistical methods for environmental monitoring (see http://www.epa.gov/lawsregs/policy/sgd/byoffice.html).

1.3   What Is EnvStats?

EnvStats is an R package for environmental statistics. It is the open-source successor to the commercial module for S-Plus$^{©}$ called EnvironmentalStats for S-Plus, which was first released in April, 1997. The EnvStats package, along with the R software environment, provides comprehensive and powerful software for environmental data analysis. EnvStats brings the major environmental statistical methods found in the literature and regulatory guidance documents into one statistical package, along with an extensive hypertext help system that explains what these methods do, *how* to use these methods, and *where* to find them in the environmental statistics literature. Also included are numerous built-in data sets from regulatory guidance documents and the environmental statistics literature. EnvStats combined with other R packages (e.g., for spatial analysis) provides the environmental scientist, statistician, researcher, and technician with tools to "get the job done!"

Because EnvStats is an R package, you automatically have access to all the features and functions of R, including powerful graphics, standard hypothesis tests, and the flexibility of a programming language. In addition, with EnvStats you can use new functions to:

- Compute several kinds of summary statistics and create summary plots to compare the distributions between groups side-by-side.
- Compute quantities (probability density functions, cumulative distribution functions, and quantiles) and random numbers associated with new probability distributions, including the extreme value distribution and the zero-modified lognormal (delta) distribution.
- Plot probability distributions so you can see how they change with the value of the distribution parameter(s).

- Estimate distribution parameters and quantiles and compute confidence intervals for commonly used probability distributions, including special methods for the lognormal and gamma distributions.
- Perform and plot the results of goodness-of-fit tests, including a new generalized goodness-of-fit test for any continuous distribution.
- Compute optimal Box-Cox data transformations.
- Compute parametric and non-parametric prediction intervals, simultaneous prediction intervals, and tolerance intervals.
- Perform additional hypothesis tests not already part of R, including non-parametric estimation and tests for seasonal trend, Fisher's one-sample randomization (permutation) test for location, the quantile test to detect a shift in the tail of one population relative to another, two-sample linear rank tests, and the test for serial correlation based on the von Neumann rank test.
- Perform power and sample size computations and create associated plots for sampling designs based on confidence intervals, hypothesis tests, prediction intervals, and tolerance intervals.
- Perform calibration based on a machine signal to determine decision and detection limits, and report estimated concentrations along with confidence intervals.
- Analyze singly and multiply censored (less-than-detection-limit) data with empirical cdf and Q-Q plots, parameter/quantile estimation and confidence intervals, prediction and tolerance intervals, goodness-of-fit tests, optimal Box-Cox transformations, and two-sample rank tests.
- Perform probabilistic risk assessment.
- Reproduce specific examples in EPA guidance documents by using built-in data sets from these documents and companion scripts.

1.4 Intended Audience and Users

EnvStats and this user's manual are intended for anyone who has to make sense of environmental data, including statisticians, environmental scientists, hydrologists, soil scientists, atmospheric scientists, geochemists, environmental engineers and consultants, hazardous and solid waste site managers, and regulatory agency analysts and enforcement officers. Some parts of EnvStats incorporate statistical methods that have appeared in the environmental literature but are not commonly found in any statistical software package. Some parts are specifically aimed at users who are required to collect and analyze environmental monitoring data in order to comply with federal and state Superfund, RCRA, CERCLA, and Subtitle D regulations for environmental monitoring at hazardous and solid waste sites. All of the functions in EnvStats, however, are useful to anyone who needs to analyze environmental data.

EnvStats is an R package. In order to use it, you need to have R installed on your system (see www.r-project.org) and know how to perform basic operations in R, such as using the command and script windows, reading data into R, and creating basic data objects (e.g., data frames). The R documentation provides information on using R, and there are several excellent introductory books on R (e.g., Allerhand 2011; Ekstrøm 2012; Zuur et al. 2009) as well as books on how to use R for specific kinds of analyses (see for example www.springer.com/series/6991). In addition, you need to have a basic knowledge of probability and statistics. While this user's manual provides *brief* explanations of various topics in environmental statistics, a companion textbook, *Environmental Statistics with R* (Millard et al. 2014), provides more details.

1.5 System Requirements

Because EnvStats is an R package, it runs under every operating system that R runs under, including Windows, UNIX-like, and MacOS X.

1.6 Installing EnvStats

To download and install the EnvStats package, the most up to date information is available on www.probstatinfo.com. This includes how to find EnvStats on the Comprehensive R Archive Network (CRAN).

1.7 Starting EnvStats

To start EnvStats, you must have already started R. You can load the package either from the command window or, if you are running R under Windows, from the menu.

- To load the package from the command window, type

```
> library(EnvStats)
```

- To load the package from the menu, on the R menu bar make the following selections: **Packages>Load package…**. This brings up a dialog box listing all the packages you currently have installed. Select **EnvStats** and click **OK**.

Loading the package attaches the library of EnvStats functions to the second position in your search list (type `search()` at the R prompt to see the listing of directories on the search list).

Note: Some of the functions in EnvStats mask built-in R functions. The masked functions are modified versions of the built-in functions and have been created to support the other functions in EnvStats, but the modifications should not affect normal use of R. If you experience unexpected behavior of R after loading EnvStats, try unloading the package (see the Sect. 1.10 below). All of the functions in EnvStats, whether they mask built-in functions or not, are described in the help system.

1.8 Getting Help and Using Companion Scripts

To get help for a specific topic, function or dataset, use the ? operator or the `help` function, just as you do for other help topics in R. For example, type

```
> ?EnvStats
```

or

```
> help(EnvStats)
```

to display the general help file for the EnvStats package, type

```
> ?pdfPlot
```

or

```
> help(pdfPlot)
```

to display the help file for the `pdfPlot` function, and type

```
> ?EPA.94b.tccb.df
```

or

```
> help(EPA.94b.tccb.df)
```

to display the help file for the data frame `EPA.94b.tccb.df`. For help files with an associated alias of more than one word, you'll need to enclose the words in quotes. For example, typing

```
> ?"Functions by Category"
```

will bring up a help file with a hyperlink list of EnvStats functions by category. Running the command

```
> help(package="EnvStats")
```

will bring up a window with basic information about the EnvStats package, including the version number and a list of functions and datasets. All function names start with a lower case letter, and all data object names start with an upper case letter. Typing

```
> newsEnvStats()
```

will bring up a window with the latest information about the current version of EnvStats.

There is also a companion PDF file **EnvStats-maual.PDF** containing a listing of all the help files. This file is located in the **doc** subdirectory of the directory where the EnvStats package was installed. For example, if you installed R under Windows, this file might be located in the directory **C:\Program Files\R-*.**.**\library\EnvStats\doc** where *.**.* denotes the version of R you are using (e.g., 3.0.1) or in the directory **C:\Users*Name*\Documents\R\win-library*.**.**\EnvStats\doc** where *Name* denotes your user name on the Windows operating system.

EnvStats comes with sets of companion scripts, located in the **scripts** subdirectory of the directory where EnvStats was installed. The scripts located in the directory **Manual** let you reproduce the output and figures shown in this user's manual. The scripts in the other directories let you reproduce examples from various US EPA guidance documents.

1.9   A Note About Examples and Masking

Some of the examples in this user's manual ask you to attach a particular data frame in order to access the variables within the data frame directly. For example, typing the command:

```
> attach(EPA.94b.tccb.df)
```

will attach the data frame EPA.94b.tccb.df to your search list. The names of the variables in this data frame are:

```
> names(EPA.94b.tccb.df)
```

```
[1] "TcCB.orig" "TcCB"       "Censored"   "Area"
```

If you already have a data object in your working directory with one of these names (e.g., Area), then it will mask the variable that is part of the attached data frame. In this case you will probably not get the same results as those shown in this user's manual. Therefore, it is recommended that when you are typing in the examples in this user's manual or running the companion scripts, you do so from a working directory that has no pre-existing data objects.

If you do attach a data object in order to access variables from it directly, it is good practice to detach it after you are through with it. For example, typing the command:

```
> detach("EPA.94b.tccb.df")
```

will detach the data frame EPA.94b.tccb.df from your search list. In this user's manual, we do not explicitly show commands for detaching data objects; however the companion scripts include commands for detaching data objects.

To avoid problems with masking in your day-to-day work, apart from always starting with a working directory that has no pre-existing objects in it (not likely since it is common practice to keep objects for a project in a specific working directory), there are at least three things you can do:

1. For functions that take a data argument, specify the data source in this way. For example, typing:

    ```
    > stripChart(TcCB ~ Area, data = EPA.94b.tccb.df)
    ```

    will create strip charts of the TcCB data by area by using the data in the column labeled TcCB in the data frame EPA.94b.tccb.df and using the data in the column Area to specify the area.

2. For functions that don't take a data argument, you can use the R with function. For example, typing

    ```
    > with(EPA.94b.tccb.df, summary(TcCB))
    ```

    will give you summary statistics for the column labeled TcCB in the data frame EPA.94b.tccb.df.

3. Access the variables directly using the [or $ operators. For example, typing:

    ```
    > summary(EPA.94b.tccb.df$TcCB)
    ```

    will give you summary statistics for the column labeled TcCB in the data frame EPA.94b.tccb.df.

1.10 Unloading EnvStats

To remove the EnvStats package from your R session, type the following command at the R prompt:

```
> detach(package:EnvStats)
```

1.11 A Tutorial

This section highlights some of the major features of EnvStats. There are several ways to use this section. If you are fairly new to R, you may want to briefly skim this section to get an idea of what you can do in EnvStats, and then come back later after you have read the other chapters of this manual. If you have used R for a long time and have just installed EnvStats, you may want to follow this tutorial in depth right now to get acquainted with some of the features available in this R package. Throughout this section we assume you have started R and also have loaded EnvStats.

1.11.1 The TcCB Data

The guidance document *Statistical Methods for Evaluating the Attainment of Cleanup Standards, Volume 3: Reference-Based Standards for Soils and Solid Media* (USEPA 1994b, pp. 6.22–6.25) contains 124 measures of 1,2,3,4-Tetrachlorobenzene (TcCB) concentrations (ppb) from soil samples at a Reference site and a Cleanup area (Table 1.1). There are 47 observations from the Reference site and 77 in the Cleanup area. These data are stored in the data frame EPA.94b.tccb.df. There is one observation coded as "ND" in this data set as presented in the guidance document. In the data frame EPA.94b.tccb.df this observation is treated as censored at an assumed detection limit of 0.09 ppb (the smallest observed value). For the purposes of this tutorial, we'll ignore the fact that this observation is censored and assumed it has an observed value of 0.09. See Chap. 8 for more information on dealing with censored data.

Reference area				Cleanup area						
0.22	0.23	0.26	0.27	<0.09	0.09	0.09	0.12	0.12	0.14	0.16
0.28	0.28	0.29	0.33	0.17	0.17	0.17	0.18	0.19	0.20	0.20
0.34	0.35	0.38	0.39	0.21	0.21	0.22	0.22	0.22	0.23	0.24
0.39	0.42	0.42	0.43	0.25	0.25	0.25	0.25	0.26	0.28	0.28
0.45	0.46	0.48	0.50	0.29	0.31	0.33	0.33	0.33	0.34	0.37
0.50	0.51	0.52	0.54	0.38	0.39	0.40	0.43	0.43	0.47	0.48
0.56	0.56	0.57	0.57	0.48	0.49	0.51	0.51	0.54	0.60	0.61
0.60	0.62	0.63	0.67	0.62	0.75	0.82	0.85	0.92	0.94	1.05
0.69	0.72	0.74	0.76	1.10	1.10	1.19	1.22	1.33	1.39	1.39
0.79	0.81	0.82	0.84	1.52	1.53	1.73	2.35	2.46	2.59	2.61
0.89	1.11	1.13	1.14	3.06	3.29	5.56	6.61	18.40	51.97	
1.14	1.20	1.33		168.64						

Table 1.1 1,2,3,4-Tetrachlorobenzene (TcCB) concentrations (ppb) from soil samples

To look at the raw data, after loading **EnvStats**, type EPA.94b.tccb.df at the R command prompt:

```
> library(EnvStats)

> EPA.94b.tccb.df

    TcCB.orig   TcCB Censored      Area
1        0.22   0.22    FALSE Reference
2        0.23   0.23    FALSE Reference
...
48      <0.09   0.09     TRUE   Cleanup
...
123     51.97  51.97    FALSE   Cleanup
124    168.64 168.64    FALSE   Cleanup
```

If you just want to get a feel for the data and don't need to look at all of the rows, you can use the R head function to look at just the first few rows:

```
> head(EPA.94b.tccb.df)
```

```
  TcCB.orig TcCB Censored      Area
1      0.22 0.22    FALSE Reference
2      0.23 0.23    FALSE Reference
3      0.26 0.26    FALSE Reference
4      0.27 0.27    FALSE Reference
5      0.28 0.28    FALSE Reference
6      0.28 0.28    FALSE Reference
```

For the remainder of this tutorial, we will assume that you have attached the data frame EPA.94b.tccb.df to your search list with the following command:

```
> attach(EPA.94b.tccb.df)
```

1.11.2 Computing Summary Statistics

There are two different functions in EnvStats for producing summary statistics. The function summaryStats produces commonly reported summary statistics while the function summaryFull provides a much more extensive set:

```
> summaryStats(TcCB ~ Area, data = EPA.94b.tccb.df)
```

```
           N Mean   SD Median Min   Max
Cleanup   77  3.9 20.0    0.4 0.1 168.6
Reference 47  0.6  0.3    0.5 0.2   1.3
```

```
> summaryFull(TcCB ~ Area, data = EPA.94b.tccb.df)
```

	Cleanup	Reference
Sample Size:	77	47
Mean:	3.915	0.5985
Median:	0.43	0.54
10% Trimmed Mean:	0.6846	0.5728
Geometric Mean:	0.5784	0.5382
Skew:	7.717	0.9019
Kurtosis:	62.67	0.132
Min:	0.09	0.22
Max:	168.6	1.33
Range:	168.5	1.11
1st Quartile:	0.23	0.39
3rd Quartile:	1.1	0.75
Standard Deviation:	20.02	0.2836
Geometric Standard Deviation:	3.898	1.597
Interquartile Range:	0.87	0.36
Median Absolute Deviation:	0.3558	0.2669
Coefficient of Variation:	5.112	0.4739

These summary statistics indicate that the observations for the Cleanup area are extremely skewed to the right. This may be indicative of residual contamination.

1.11.3 Looking at the TcCB Data

Figure 1.1 shows one-dimensional scatterplots (also called strip plots or strip charts) of the log-transformed TcCB data by area, along with confidence intervals for the means, created with the EnvStats function `stripChart` (a modification of the R function `stripchart`):

```
> stripChart(log(TcCB) ~ Area, data = EPA.94b.tccb.df,
    col = c("red", "blue"), ylab = "Log [ TcCB (ppb) ]")
```

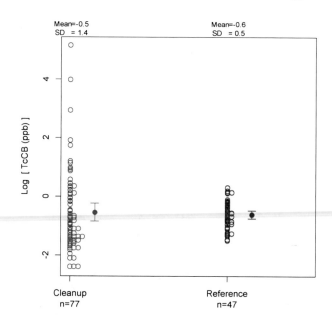

Fig. 1.1 One-dimensional scatterplots, along with 95 % confidence intervals for the mean, comparing log-transformed TcCB concentrations at Reference and Cleanup areas

Figure 1.2 shows the associated histograms produced with the R function `hist`:

```
> par(mfrow = c(2, 1))

> hist(log(TcCB[Area == "Reference"]), xlim = c(-4, 6),
    col = "blue", xlab = "log [ TcCB (ppb) ]",
    ylab = "Number of Observations", main = "Reference Area")

> hist(log(TcCB[Area == "Cleanup"]), xlim = c(-4, 6),
    nclass = 30, col = "red", xlab = "log [ TcCB (ppb) ]",
    ylab = "Number of Observations", main = "Cleanup Area")
```

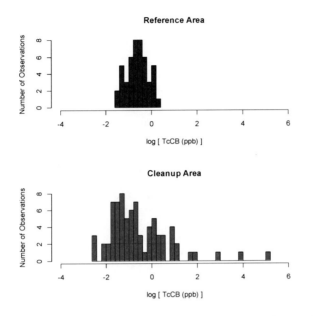

Fig. 1.2 Histograms comparing log-transformed TcCB concentrations at Reference and Cleanup areas

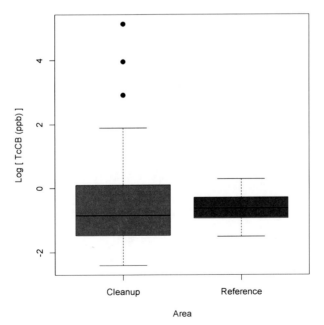

Fig. 1.3 Boxplots comparing log-transformed TcCB concentrations at Reference and Cleanup areas

Figure 1.3 shows side-by-side boxplots produced with the R function `boxplot`:

```
> boxplot(log(TcCB) ~ Area, col = c("red", "blue"),
    pars = list(outpch = 16), xlab = "Area",
    ylab = "Log [ TcCB (ppb) ]")
```

We see in these plots that most of the observations in the Cleanup area are comparable to (or even smaller than) the observations in the Reference area, but there are a few very large "outliers" in the Cleanup area. As previously stated, this may be indicative of residual contamination that was missed during the cleanup process.

1.11.4 Quantile (Empirical CDF) Plots

Figure 1.4 shows the quantile plot, also called the empirical cumulative distribution function (cdf) plot, for the Reference area TcCB data. It was created with the EnvStats function `ecdfPlot`:

```
> ecdfPlot(TcCB[Area == "Reference"], xlab = "TcCB (ppb)")
```

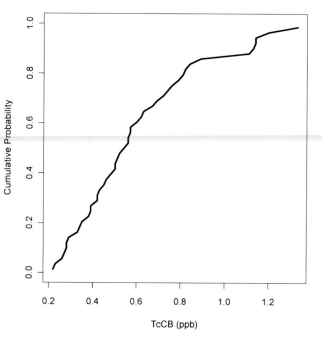

Fig. 1.4 Quantile plot of Reference area TcCB data

You can easily pick out the median as about 0.55 ppb and the quartiles as about 0.4 and 0.75 ppb (compare these numbers to the ones listed in Sect. 1.11.2). You can also see that the quantile plot quickly rises, then pretty much levels off after about 0.8 ppb, which indicates that the data are skewed to the right (see the histogram for the Reference area data in Fig. 1.2).

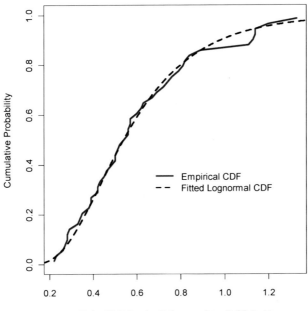

Fig. 1.5 Empirical cdf of Reference area TcCB data compared to a lognormal cdf

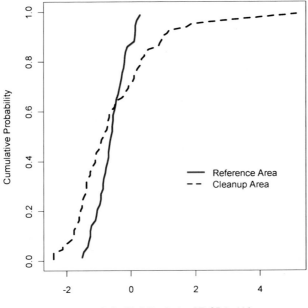

Fig. 1.6 Quantile plots comparing log-transformed TcCB data at the Reference and Cleanup areas

Figure 1.5 shows the quantile plot with a fitted lognormal distribution:

```
> cdfCompare(TcCB[Area == "Reference"], dist = "lnorm",
    xlab = "Order Statistics for Reference Area TcCB (ppb)")

> legend(0.65, 0.4, legend = c("Empirical CDF",
    "Fitted Lognormal CDF"), lty = 1:2, col = c(4, 1),
    lwd = 3, bty = "n")
```

We see that the lognormal distribution appears to fit these data quite well. Figure 1.6 compares the empirical cdf for the Reference area with the empirical cdf for the Cleanup area for the log-transformed TcCB data:

```
> cdfCompare(log(TcCB[Area == "Reference"]),
    log(TcCB[Area == "Cleanup"]),
    xlab = "Order Statistics for log [ TcCB (ppb) ]")

> legend(1.5, 0.4, legend = c("Reference Area",
    "Cleanup Area"), lty = 1:2, col = c(4, 1), lwd = 3,
    bty = "n")
```

As we saw with the histograms and boxplots, the Cleanup area has quite a few extreme values compared to the Reference area.

1.11.5 Assessing Goodness-of-Fit with Quantile-Quantile Plots

Figure 1.7 displays the normal Q-Q plot for the log-transformed Reference area TcCB data (i.e., we are assuming these data come from a lognormal distribution), along with a fitted least squares line.

```
> qqPlot(TcCB[Area == "Reference"], dist = "lnorm",
    add.line = TRUE, points.col = "blue",
    ylab="Quantiles of log [ TcCB (ppb) ]")
```

Figure 1.8 displays the corresponding Tukey mean-difference Q-Q plot.

```
> qqPlot(TcCB[Area == "Reference"], dist = "lnorm",
    plot.type = "Tukey", estimate.params = TRUE,
    add.line = TRUE, points.col = "blue")
```

As we saw with the quantile plot, the lognormal model appears to be a fairly good fit to these data.

Some EPA guidance documents (e.g., Singh et al. 2002; Singh et al. 2010a,b) discourage using the assumption of a lognormal distribution and recommend instead assessing whether the data appear to fit a gamma distribution. Figure 1.9 displays the gamma Q-Q plot for the Reference area TcCB data:

```
> qqPlot(TcCB[Area == "Reference"], dist = "gamma",
    estimate.params = TRUE, digits = 2, add.line = TRUE,
    points.col = "blue", ylab="Quantiles of TcCB (ppb)")
```

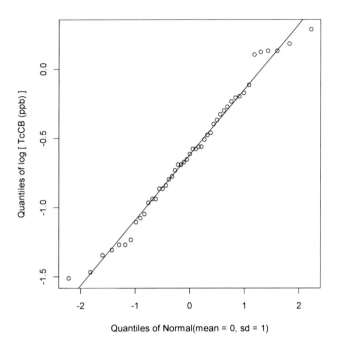

Fig. 1.7 Normal Q-Q plot for the log-transformed Reference area TcCB data

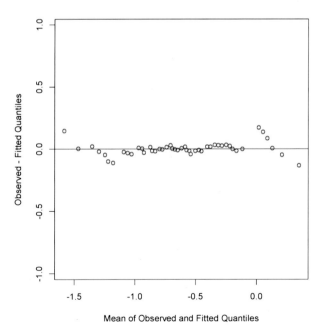

Fig. 1.8 Tukey mean-difference Q-Q plot for the Reference area TcCB data fitted to a lognormal distribution

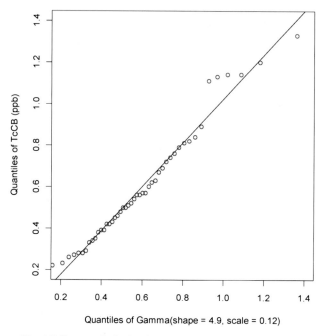

Fig. 1.9 Gamma Q-Q plot for the Reference area TcCB data

The gamma model also appears to be a fairly good fit to these data.

1.11.6 Estimating Distribution Parameters

In EnvStats you can estimate parameters for several parametric distributions. For example, for the lognormal distribution you can estimate the mean and standard deviation based on the log-transformed data, or you can estimate the mean and coefficient of variation based on the original data. For either parameterization, you can compute a confidence interval for the mean. Here are the results for the log-transformed Reference area TcCB data:

```
> elnorm(TcCB[Area == "Reference"], ci = TRUE)

Results of Distribution Parameter Estimation
--------------------------------------------------

Assumed Distribution:            Lognormal

Estimated Parameter(s):          meanlog = -0.6195712
                                 sdlog   =  0.4679530

Estimation Method:               mvue

Data:                            TcCB[Area == "Reference"]
```

```
Sample Size:                        47

Confidence Interval for:            meanlog

Confidence Interval Method:         Exact

Confidence Interval Type:           two-sided

Confidence Level:                   95%

Confidence Interval:                LCL = -0.7569673
                                    UCL = -0.4821751
```

and here are the results for the original Reference area TcCB data:

```
> elnormAlt(TcCB[Area == "Reference"], ci = TRUE)
Results of Distribution Parameter Estimation
--------------------------------------------

Assumed Distribution:               Lognormal

Estimated Parameter(s):             mean = 0.5989072
                                    cv   = 0.4899539

Estimation Method:                  mvue

Data:                               TcCB[Area == "Reference"]

Sample Size:                        47

Confidence Interval for:            mean

Confidence Interval Method:         Land
Confidence Interval Type:           two-sided

Confidence Level:                   95%

Confidence Interval:                LCL = 0.5243787
                                    UCL = 0.7016992
```

If we want to assume a gamma distribution, we can use the egamma function, which estimates the shape and scale parameters and gives a confidence interval for the mean:

```
> egamma(TcCB[Area == "Reference"], ci = TRUE)
```

```
Results of Distribution Parameter Estimation
--------------------------------------------

Assumed Distribution:            Gamma

Estimated Parameter(s):          shape = 4.8659316
                                 scale = 0.1230002

Estimation Method:               mle

Data:                            TcCB[Area == "Reference"]

Sample Size:                     47

Confidence Interval for:         mean

Confidence Interval Method:      Optimum Power
                                 Normal Approximation
                                 of Kulkarni & Powar (2010)
                                 using mle of 'shape'

Confidence Interval Type:        two-sided

Confidence Level:                95%

Confidence Interval:             LCL = 0.5196677
                                 UCL = 0.6844993
```

Or, we can use the egammaAlt function, which estimates the mean and coefficient of variation, and gives a confidence interval for the mean:

```
> egammaAlt(TcCB[Area == "Reference"], ci = TRUE)

Results of Distribution Parameter Estimation
--------------------------------------------

Assumed Distribution:            Gamma

Estimated Parameter(s):          mean = 0.5985106
                                 cv   = 0.4533326

Estimation Method:               mle of 'shape'

Data:                            TcCB[Area == "Reference"]
```

```
Sample Size:                   47

Confidence Interval for:       mean

Confidence Interval Method:    Optimum Power
                               Normal Approximation
                               of Kulkarni & Powar (2010)
                               using mle of 'shape'

Confidence Interval Type:      two-sided

Confidence Level:              95%

Confidence Interval:           LCL = 0.5196677
                               UCL = 0.6844993
```

You can see that in this case the 95 % confidence intervals for the mean based on the lognormal distribution and based on the gamma distribution are nearly identical.

1.11.7 Testing for Goodness of Fit

EnvStats contains several new or modified R functions for testing goodness of fit. Here we will use the Shapiro-Wilk test to test the adequacy of a lognormal model and then a gamma model for the Reference area TcCB data (Figs. 1.10 and 1.11).

```
> TcCB.ref <- TcCB[Area == "Reference"]

> sw.lnorm <- gofTest(TcCB.ref, dist = "lnorm")

> sw.lnorm

Results of Goodness-of-Fit Test
-------------------------------

Test Method:                   Shapiro-Wilk GOF

Hypothesized Distribution:     Lognormal

Estimated Parameter(s):        meanlog = -0.6195712
                               sdlog   =  0.4679530

Estimation Method:             mvue

Data:                          TcCB.ref

Sample Size:                   47
```

```
Test Statistic:                    W = 0.978638

Test Statistic Parameter:          n = 47

P-value:                           0.5371935

Alternative Hypothesis:            True cdf does not equal the
                                   Lognormal Distribution.
```

```
> plot(sw.lnorm, digits = 3)
```

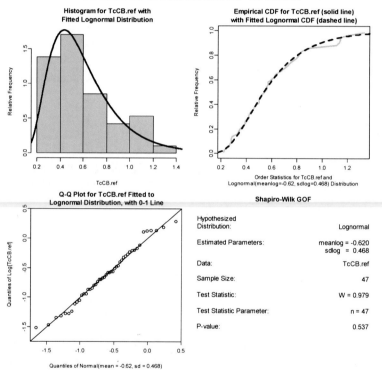

Fig. 1.10 Summary plots of Shapiro-Wilk goodness-of-fit test for lognormal distribution for Reference area TcCB data

```
> sw.gamma <- gofTest(TcCB.ref, dist = "gamma")

> plot(sw.gamma, digits = 3)
```

Goodness-of-Fit Results for TcCB.ref

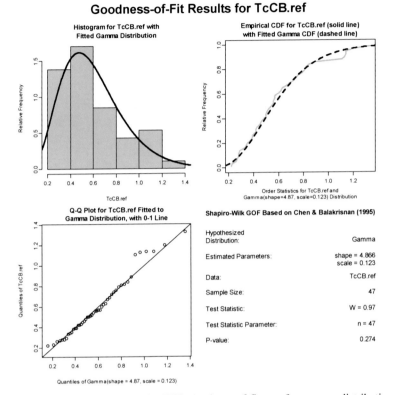

Fig. 1.11 Summary plots of Shapiro-Wilk goodness-of-fit test for gamma distribution for Reference area TcCB data

```
> sw.gamma

Results of Goodness-of-Fit Test
-------------------------------


Test Method:               Shapiro-Wilk GOF using Probabilities
                           Based on Fitted Distribution

Hypothesized Distribution:  Gamma

Estimated Parameter(s):    shape = 4.8659316
                           scale = 0.1230002

Estimation Method:         mle

Data:                      TcCB.ref
```

```
Sample Size:                        47

Test Statistic:                     W = 0.9703805

Test Statistic Parameter:           n = 47

P-value:                            0.2738988

Alternative Hypothesis:             True cdf does not equal the
                                    Gamma Distribution.
```

The goodness-of-fit tests show that both the lognormal and gamma distributions appear to fit the Reference area TcCB data.

1.11.8 Estimating Quantiles and Computing Confidence Limits

EnvStats contains functions for estimating quantiles and optionally constructing confidence limits for the quantiles. Here we will estimate the 90th percentile of the distribution of the Reference area TcCB, assuming the true distribution is a lognormal distribution, and compute a 95 % confidence interval for this 90th percentile.

```
> eqlnorm(TcCB[Area == "Reference"], p = 0.9, ci = TRUE)

Results of Distribution Parameter Estimation
--------------------------------------------------
```

```
Assumed Distribution:               Lognormal

Estimated Parameter(s):             meanlog = -0.6195712
                                    sdlog   =  0.4679530
Estimation Method:                  mvue

Estimated Quantile(s):              90'th %ile = 0.9803307

Quantile Estimation Method:         qmle

Data:                               TcCB[Area == "Reference"]

Sample Size:                        47

Confidence Interval for:            90'th %ile

Confidence Interval Method:         Exact

Confidence Interval Type:           two-sided
```

```
Confidence Level:              95%

Confidence Interval:           LCL = 0.8358791
                               UCL = 1.2154977
```

1.11.9 Comparing Two Distributions Using Nonparametric Tests

EnvStats contains functions for performing general two-sample linear rank tests (to test for a shift in location) and a special quantile test that tests for a shift in the tail of one of the distributions. In this example we will compare the Reference and Cleanup area TcCB data. Here are the results for the Wilcoxon rank sum test:

```
> twoSampleLinearRankTest(TcCB[Area == "Cleanup"],
    TcCB[Area == "Reference"], alternative = "greater")

Results of Hypothesis Test
--------------------------


Null Hypothesis:           Fy(t) = Fx(t)

Alternative Hypothesis:    Fy(t) > Fx(t) for at least one t

Test Name:                 Two-Sample Linear Rank Test:
                           Wilcoxon Rank Sum Test
                           Based on Normal Approximation

Data:                      x = TcCB[Area == "Cleanup"]
                           y = TcCB[Area == "Reference"]

Sample Sizes:              nx = 77
                           ny = 47

Test Statistic:            z = -1.171872

P-value:                   0.8793758
```

and here are the results for the quantile test:

```
> quantileTest(TcCB[Area == "Cleanup"],
    TcCB[Area == "Reference"], alternative = "greater",
    target.r = 9)

Results of Hypothesis Test
--------------------------


Null Hypothesis:               e = 0
```

```
Alternative Hypothesis:        Tail of Fx Shifted to Right of
                               Tail of Fy.
                               0 < e <= 1, where
                               Fx(t)  =  (1-e)*Fy(t) + e*Fz(t),
                               Fz(t)  <= Fy(t) for all t,
                               and Fy != Fz

Test Name:                     Quantile Test

Data:                          x = TcCB[Area == "Cleanup"]
                               y = TcCB[Area == "Reference"]

Sample Sizes:                  nx = 77
                               ny = 47

Test Statistics:               k (# x obs of r largest)  = 9
                               r                         = 9

Test Statistic Parameters:     m               = 77.000
                               n               = 47.000
                               quantile.ub =   0.928

P-value:                       0.01136926
```

Note that the Wilcoxon rank sum test is not significant at the 0.10 level ($p = 0.88$), while the quantile test is significant at the 0.011 level. The quantile test picked up the portion of large outlying values in the Cleanup area data. Note: you can also perform the Wilcoxon rank sum test with the R function `wilcox.test`. The EnvStats function `two.sample.linear.rank.test` lets you perform other kinds of linear rank tests, including normal scores, Mood's median, and Savage scores.

1.12 Summary

- Environmental statistics is the application of statistics to environmental problems.
- EnvStats is an R package for environmental statistics. It includes several functions for creating graphs and performing statistical analyses that are commonly used in environmental statistics.
- To use EnvStats you should be familiar with the basic operation of R and have an elementary knowledge of probability and statistics.
- EnvStats has an extensive help system that includes basic explanations in English, as well as equations and references.

Chapter 2

Designing a Sampling Program

2.1 Introduction

The first and most important step of any environmental study is to design the sampling program. This chapter discusses the basics of designing a sampling program, and shows you how to use EnvStats to help you determine required sample sizes. For a more in-depth discussion of sampling design and sample size calculation, see Millard et al. (2014).

2.2 The Necessity of a Good Sampling Design

A study is only as good as the data upon which it is based. No amount of advanced, cutting-edge statistical theory and techniques can rescue a study that has produced poor quality data, not enough data, or data irrelevant to the question it was meant to answer. From the very start of an environmental study, there must be a constant dialog between the data producers (field and lab personnel, data coders, etc.), the data users (scientists and statisticians), and the ultimate decision maker (the person for whom the study was instigated in the first place). All persons involved in the study must have a clear understanding of the study objectives and the limitations associated with the chosen physical sampling and analytical (measurement) techniques before anyone can make any sense of the resulting data.

2.3 What Is a Population and What Is a Sample?

In everyday language, the word "population" refers to all the people or organisms contained within a specific country, area, region, etc. When we talk about the population of the United States, we usually mean something like "the total number of people who currently reside in the U.S."

In the field of statistics, however, the term *population* is defined operationally by the question we ask: it is the entire collection of measurements about which we want to make a statement (Zar 2010; Berthouex and Brown 2002; Gilbert 1987).

For example, if the question is "What is the concentration of dissolved oxygen in this stream?", the question must be further refined until a suitable population can be defined: "What is the average concentration of dissolved oxygen in a particular section of a stream at a depth of 0.5 m over a particular 3-day period?" In this case, the population is the set of all possible measurements of dissolved oxygen in

S.P. Millard, *EnvStats: An R Package for Environmental Statistics*,
DOI 10.1007/978-1-4614-8456-1_2, © Springer Science+Business Media New York 2013

that section of the stream at 0.5 m within that time period. The section of the stream, the time period, the method of taking water samples, and the method of measuring dissolved oxygen all define the population.

A *sample* is defined as some subset of a population (Zar 2010; Berthouex and Brown 2002; Gilbert 1987). If the sample contains all the elements of the population, it is called a *census*. Usually, a population is too large to take a census, so a portion of the population is sampled. The statistical definition of the word sample (a selection of individual population members) should not be confused with the more common meaning of a *physical sample* of soil (e.g., 10 g of soil), water (e.g., 5 ml of water), air (e.g., 20 cc of air), etc.

2.4   Random Versus Judgment Sampling

Judgment sampling involves subjective selection of the population units by an individual or group of individuals (Gilbert 1987). For example, the number of samples and sampling locations might be determined based on expert opinion or historical information. Sometimes, public opinion might play a role and samples need to be collected from areas known to be highly polluted. The uncertainty inherent in the results of a judgment sample cannot be quantified and statistical methods cannot be applied to judgment samples. Judgment sampling does *not* refer to using prior information and the knowledge of experts to define the area of concern, define the population, or plan the study. Gilbert (1987) also describes "haphazard" sampling, which is a kind of judgment sampling with the attitude that "any sample will do" and can lead to "convenience" sampling, in which samples are taken in convenient places at convenient times.

Probability sampling or *random sampling* involves using a random mechanism to select samples from the population (Gilbert 1987). All statistical methods used to quantify uncertainty assume some form of random sampling has been used to obtain a sample. At the simplest level, a simple random sample is used in which each member of the population has an equal chance of being chosen, and the selection of any member of the population does not influence the selection of any other member. Other probability sampling methods include stratified random sampling, composite sampling, and ranked set sampling.

2.5   Common Mistakes in Environmental Studies

The most common mistakes that occur in environmental studies include the following:

- **Using Judgment Sampling to Obtain Samples.** When judgment sampling is used to obtain samples, there is no way to quantify the precision and bias of any type of estimate computed from these samples.
- **Lack of Samples from Proper Control Populations**. If one of the objectives of an environmental study is to determine the effects of a pollutant on some specified population, then the sampling design must

include samples from a proper control population. This is a basic tenet of the scientific method. If control populations were not sampled, there is no way to know whether the observed effect was really due to the hypothesized cause, or whether it would have occurred anyway.

- **Failing to Randomize over Potentially Influential Factors.** An enormous number of factors can influence the final measure associated with a single sampling unit, including the person doing the sampling, the device used to collect the sample, the weather and field conditions when the sample was collected, the method used to analyze the sample, the laboratory to which the sample was sent, etc. A good sampling design controls for as many potentially influencing factors as possible, and randomizes over the factors that cannot be controlled. For example, if data are collected from two sites, and two laboratories are used to analyze the results, you should not send all the samples from site 1 to laboratory A and all the samples from site 2 to laboratory B, but rather send samples collected at each site to each of the laboratories.

- **Collecting Too Few Samples to Have a High Degree of Confidence in the Results.** The ultimate goal of an environmental study is to answer one or more basic questions. These questions should be stated in terms of hypotheses that can be tested using statistical procedures, as well as what constitutes an important scientific effect since statistically significant effects are not always scientifically important. In this case, you can determine the probability of rejecting the null hypothesis when in fact it is true (a Type I error), and the probability of not rejecting the null hypothesis when in fact it is false (a Type II error). Usually, the Type I error is set in advance, and the probability of correctly rejecting the null hypothesis when in fact it is false (the power), or the width of a confidence, prediction, or tolerance interval, is calculated for various sample sizes and assumed amounts of variability. Too often, this step of determining power and/or interval width versus sample size is neglected, resulting in a study from which no conclusions can be drawn with any great degree of confidence.

2.6 The Data Quality Objectives Process

The Data Quality Objectives (DQO) process is a systematic planning tool based on the scientific method that has been developed by the U.S. Environmental Protection Agency (USEPA 2006b). The DQO process provides an easy-to-follow, step-by-step approach to decision-making in the face of uncertainty. Each step focuses on a specific aspect of the decision-making process. Data Quality Objectives are the qualitative and quantitative statements that:

- Clarify the study objective.
- Define the most appropriate type of data to collect.
- Determine the most appropriate conditions under which to collect the data.

- Specify acceptable levels of decision errors that will be used as the basis for establishing the quantity and quality of data needed to support the decision.

The seven steps in the DQO process are: (1) state the problem, (2) identify the goals of the study, (3) identify information inputs, (4) define boundaries of the study, (5) develop the analytic approach, (6) specify performance or acceptance criteria, and (7) develop the plan for obtaining the data (see Millard et al. 2014, for more details). Steps 5 and 6 involve deciding what statistical methods you will use and trading off limits on Type I and Type II errors and sample size.

2.7   Power and Sample Size Calculations

EnvStats contains several functions to assist you in determining how many samples you need for a given degree of confidence in the results of a sampling program (see the help file *Power and Sample Size*). These functions are based on the ideas of confidence intervals, prediction intervals, tolerance intervals, and hypothesis tests. If you are unfamiliar with these concepts, please see Millard et al. (2014).

A very important point to remember is that no matter what you come up with for estimates of required sample sizes, it is always a good idea to assume you will lose some percentage of your observations due to sample loss, sample contamination, database issues, etc.

2.8   Sample Size for Confidence Intervals

Table 2.1 lists the functions available in EnvStats for computing required sample sizes, half-widths, and confidence levels associated with a confidence interval. For the normal and binomial distributions, you can compute the half-width of the confidence interval given the user-specified sample size, compute the required sample size given the user-specified half-width, and plot the relationship between sample size and half-width. For a nonparametric confidence interval for a percentile, you can compute the required sample size for a specified confidence level, compute the confidence level associated with a given sample size, and plot the relationship between sample size and confidence level. Chapter 5 gives more details on computing confidence intervals once you have your data.

Bacchetti (2010) presents strong arguments against the current convention in scientific research for computing sample size that is based on formulas that use a fixed Type I error (usually 5 %) and a fixed minimal power (often 80 %) without regard to costs. He notes that a key input to these formulas is a measure of variability (usually a standard deviation) that is difficult to measure accurately "unless there is so much preliminary data that the study isn't really needed." Also, study designers often avoid defining what a scientifically meaningful difference is by presenting sample size results in terms of the effect size (i.e., the difference of interest divided by the elusive standard deviation). Bacchetti (2010) encourages study designers to use simple tables in a sensitivity analysis to see what results of a study may look like for low, moderate, and high rates of variability and large, intermediate, and no underlying differences in the populations or processes being studied.

Distribution	Function	Output
Normal	`ciTableMean`	Confidence intervals for mean of normal distribution, or difference between two means, following Bacchetti (2010)
	`ciNormHalfWidth`	Half-width of confidence interval for mean of normal distribution or difference between two means
	`ciNormN`	Required sample size for specified half-width of confidence interval for mean of normal distribution or difference between two means
	`plotCiNormDesign`	Plots for sampling design based on confidence interval for mean of normal distribution or difference between two means
Binomial	`ciTableProp`	Confidence intervals for binomial proportion, or difference between two proportions, following Bacchetti (2010)
	`ciBinomHalfWidth`	Half-width of confidence interval for binomial proportion or difference between two proportions
	`ciBinomN`	Required sample size for specified half-width of confidence interval for binomial proportion or difference between two proportions
	`plotCiBinomDesign`	Plots for sampling design based on confidence interval for binomial proportion or difference between two proportions
Nonparametric	`ciNparConfLevel`	Confidence level of confidence interval for a percentile, given the sample size
	`ciNparN`	Required sample size for specified confidence level of confidence interval for a percentile
	`plotCiNparDesign`	Plots for sampling design based on confidence interval for a percentile

Table 2.1 Sample size functions for confidence intervals

2.8.1 Confidence Interval for the Mean of a Normal Distribution

The EnvStats function `ciTableMean` produces a table similar to Table 1 of Bacchetti (2010) for looking at how the confidence interval for the mean of a normal distribution or the difference between two means varies with various levels of variability and the value of the estimated mean or difference between two means, given the sample size and confidence level. The EnvStats function `ciNormHalfWidth` computes the half-width associated with the confidence interval, given the sample size, estimated standard deviation, and confidence level. The function `ciNormN` computes the sample size required to achieve a specified half-width, given the estimated standard deviation and confidence level. The function `plotCiNormDesign` plots the relationships between sample size, half-width, estimated standard deviation, and confidence level.

The data frame `EPA.09.Ex.16.1.sulfate.df` contains sulfate concentrations (ppm) at one background and one downgradient well. The estimated mean and standard deviation for the background well are 536 and 27 ppm, respectively, based on a sample size of $n = 8$ quarterly samples take over 2 years. A two-sided 95 % confidence interval for this mean is [514, 559], which has a half-width of 23 ppm.

```
> EPA.09.Ex.16.1.sulfate.df

   Month Year     Well.type Sulfate.ppm
1    Jan 1995    Background         560
2    Apr 1995    Background         530
...

15   Jul 1996 Downgradient         610
16   Oct 1996 Downgradient         630

> Sulfate.back <- with(EPA.09.Ex.16.1.sulfate.df,
     Sulfate.ppm[Well.type == "Background"])

> enorm(Sulfate.back, ci = TRUE)

Results of Distribution Parameter Estimation
--------------------------------------------


Assumed Distribution:              Normal

Estimated Parameter(s):            mean = 536.25000
                                   sd   =  26.69270

Estimation Method:                 mvue

Data:                              Sulfate.back

Sample Size:                       8
```

```
Confidence Interval for:            mean

Confidence Interval Method:         Exact

Confidence Interval Type:           two-sided

Confidence Level:                   95%

Confidence Interval:                LCL = 513.9343
                                    UCL = 558.5657
```

Suppose we are planning a future study and are interested in the size of the confidence interval. Initially we plan to take eight quarterly samples taken over 2 years, as in the previous study. We could assume an estimated standard deviation of about 25 or 30 ppm, but based on the 95 % confidence interval for the variance, which is [311, 2,951] ppm, the true standard deviation may be as small as about 18 ppm or as large as about 54 ppm.

```
> enorm(Sulfate.back, ci = TRUE,
    ci.param = "variance")$interval
```

```
Confidence Interval for:            variance

Confidence Interval Method:         Exact

Confidence Interval Type:           two-sided

Confidence Level:                   95%

Confidence Interval:                LCL =  311.4703
                                    UCL = 2951.4119
```

Letting the estimated standard deviation vary from 15 to 60 ppm shows that the width of the confidence interval varies between about 13 and 50 ppm:

```
> ciNormHalfWidth(n.or.n1 = 8, sigma.hat = c(15, 30, 60))
```

```
[1] 12.54031 25.08063 50.16126
```

Assuming a standard deviation of about 30 ppm, if in a future study we take only four observations, the half-width of the confidence interval should be about 48 ppm:

```
> ciNormHalfWidth(n.or.n1 = 4, sigma.hat = 30)
```

```
[1] 47.73669
```

Also, if we want the confidence interval to have a half-width of 10 ppm, we would need to take $n = 38$ observations (i.e., quarterly samples taken over more than 9 years).

```
> ciNormN(half.width = 10, sigma.hat = 30)
```

```
[1] 38
```

Figure 2.1 displays the half-width of the confidence interval as a function of the sample size for various confidence levels, again assuming a standard deviation of about 30 ppm.

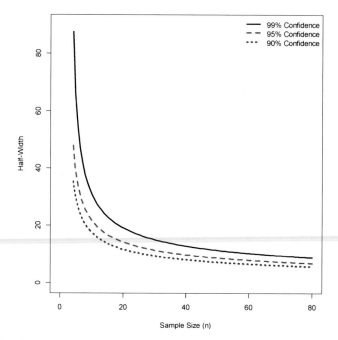

Fig. 2.1 The half-width of the confidence interval for the mean of background sulfate concentration (ppm) as a function of sample size and confidence level, assuming a standard deviation of 30 ppm

To create this plot, type these commands:

```
> plotCiNormDesign(sigma.hat = 30, range.x.var = c(4, 80),
    conf = 0.99, xlim = c(0, 80), ylim = c(0, 90), main = "")
```

```
> plotCiNormDesign(sigma.hat = 30, range.x.var = c(4, 80),
    conf = 0.95, plot.col = 2, plot.lty = 2, add = TRUE)
```

```
> plotCiNormDesign(sigma.hat = 30, range.x.var = c(4, 80),
    conf = 0.90, plot.col = 4, plot.lty = 3, add = TRUE)
```

```
> legend("topright",
    paste(c("99%", "95%", "90%"), "Confidence")
    col = c(1, 2, 4), lty = 1:3, lwd = 2, bty = "n")
```

Considering the data frame EPA.09.Ex.16.1.sulfate.df again, the estimated mean and standard deviation for the *downgradient* well are 608 and 18 ppm, respectively, based on a sample size of $n = 6$ quarterly samples. A two-sided 95 % confidence interval for the difference between this mean and the background mean is [44, 100] ppm.

```
> Sulfate.down <- with(EPA.09.Ex.16.1.sulfate.df,
    Sulfate.ppm[Well.type == "Downgradient"])

> enorm(Sulfate.down)

Results of Distribution Parameter Estimation
--------------------------------------------

Assumed Distribution:            Normal

Estimated Parameter(s):          mean = 608.33333
                                 sd   =  18.34848

Estimation Method:               mvue

Data:                            Sulfate.down

Sample Size:                     6

Number NA/NaN/Inf's:             2

> t.test(Sulfate.down, Sulfate.back,
    var.equal = TRUE)$conf.int

[1] 44.33974 99.82693
attr(,"conf.level")
[1] 0.95
```

We can use ciTableMean to look how the confidence interval for the difference between the background and downgradient means in a future study using eight quarterly samples at each well varies with assumed value of the pooled standard deviation and the observed difference between the sample means. Our current estimate of the pooled standard deviation is 24 ppm:

```
> summary(lm(Sulfate.ppm ~ Well.type,
    data = EPA.09.Ex.16.1.sulfate.df))$sigma

[1] 23.57759
```

We see that if this is overly optimistic and in our next study the pooled standard deviation is around 50 ppm, then if the observed difference between the means is 50 ppm, the lower end of the confidence interval for the difference between the two means will include 0, so we may want to increase our sample size.

```
> ciTableMean(n1 = 8, n2 = 8, diff = c(100, 50, 0),
    SD = c(15, 25, 50), digits = 0)

        Diff=100     Diff=50       Diff=0
SD=15 [ 84, 116] [ 34,   66] [-16,   16]
SD=25 [ 73, 127] [ 23,   77] [-27,   27]
SD=50 [ 46, 154] [ -4,  104] [-54,   54]
```

2.8.2   Confidence Interval for a Binomial Proportion

The EnvStats functions `ciTableProp` produces a table similar to Table 1 of Bacchetti (2010) for looking at how the confidence interval for a binomial proportion or the difference between two proportions varies with the value of the estimated proportion(s), given the sample size, confidence level, and method of computing the confidence interval. The function `ciBinomHalfWidth` computes the half-width associated with the confidence interval for the proportion or difference between two proportions, given the sample size, estimated proportion(s), confidence level, and method of computing the confidence interval. The function `ciBinomN` computes the sample size required to achieve a specified half-width, given the estimated proportion(s) and confidence level. The EnvStats function `plotCiBinomDesign` plots the relationships between sample size, half-width, estimated proportion(s), and confidence level.

The data frame `EPA.92c.benzene1.df` contains observations on benzene concentrations (ppb) in groundwater from six background wells sampled monthly for 6 months. Nondetect values are reported as "<2."

```
> EPA.92c.benzene1.df

   Benzene.orig Benzene Censored Month Well
1            <2       2     TRUE     1    1
2            <2       2     TRUE     2    1
...
35           10      10    FALSE     5    6
36           <2       2     TRUE     6    6
```

Of the 36 values, 33 are nondetects. Based on these data, the estimated probability of observing a nondetect is 92 %, and the two-sided 95 % confidence interval for the binomial proportion based on using the normal score approximation with continuity correction is [0.76, 0.98]. The half-width of this interval is 0.11, or 11 % points.

```
> with(EPA.92c.benzene1.df , ebinom(Censored, ci = TRUE))

Results of Distribution Parameter Estimation
--------------------------------------------------


Assumed Distribution:            Binomial

Estimated Parameter(s):          size = 36.0000000
                                 prob =  0.9166667

Estimation Method:               mle/mme/mvue for 'prob'

Data:                            Censored

Sample Size:                     36

Confidence Interval for:         prob

Confidence Interval Method:      Score normal approximation
                                 (With continuity correction)

Confidence Interval Type:        two-sided

Confidence Level:                95%

Confidence Interval:             LCL = 0.7640884
                                 UCL = 0.9782279
```

Suppose we are planning a future study and are interested in the size of the confidence interval. Initially we plan to take 36 samples as in the previous study. Letting the estimated percentage of nondetects vary from 75 % to 95 % shows that the width of the confidence interval varies between about 15 % and 10 % points.

```
> ciBinomHalfWidth(n.or.n1 = 36, p.hat = c(0.75, 0.85, 0.95))

$half.width
[1] 0.14907011 0.12529727 0.09523133

$n
[1] 36 36 36
$p.hat
[1] 0.7500000 0.8611111 0.9444444

$method
[1] "Score normal approximation, with continuity correction"
```

Assuming an estimated proportion of 90 %, if we take only $n = 10$ observations, the half-width of the confidence interval would be about 23 % points:

```
> ciBinomHalfWidth (n.or.n1 = 10, p.hat = 0.9)

$half.width
[1] 0.2268019

$n
[1] 10

$p.hat
[1] 0.9

$method
[1] "Score normal approximation, with continuity correction"
```

Also, if we want the confidence interval to have a half-width of 0.03 (3 % points), we would need to take $n = 319$ observations (a sample size probably not feasible for many environmental studies!).

```
> ciBinomN (half.width = 0.03, p.hat = 0.9)

$n
[1] 319

$p.hat
[1] 0.8996865

$half.width
[1] 0.03466104

$method
[1] "Score normal approximation, with continuity correction""
```

Figure 2.2 displays the half-width of the confidence interval as a function of the sample size for various confidence levels, based on using the score normal approximation with continuity correction to construct the confidence interval.

```
> plotCiBinomDesign (p.hat = 0.9, range.x.var = c (10, 200),
    conf = 0.99, xlim = c (0, 200), ylim = c (0, 0.3),
    main = "")

> plotCiBinomDesign (p.hat = 0.9, range.x.var = c (10, 200),
    conf = 0.95, plot.col = 2, plot.lty = 2, add = TRUE)

> plotCiBinomDesign (p.hat = 0.9, range.x.var = c (10, 200),
    conf = 0.90, plot.col = 4, plot.lty = 3, add = TRUE)
```

```
> legend("topright",
    paste(c("99%", "95%", "90%"), "Confidence"),
    col = c(1, 2, 4), lty = 1:3, lwd = 3, bty = "n")
```

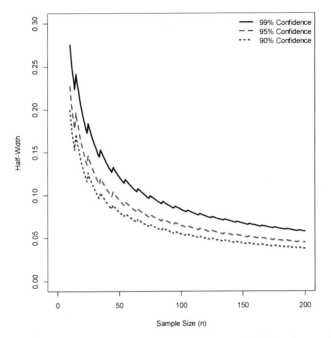

Fig. 2.2 The half-width of the confidence interval for the probability of a nondetect as a function of sample size and confidence level, assuming an estimated nondetect proportion of 90 %

If we are planning a study to compare the proportion of nondetects at a background and downgradient well, we can use `ciTableProp` to look how the confidence interval for the difference between the two proportions using say 36 quarterly samples at each well varies with the observed estimated proportions. Here we'll let the argument `p1.hat` denote the proportion of nondetects observed at the downgradient well and set this equal to 20 %, 40 % and 60 %. The argument `p2.hat.minus.p1.hat` represents the proportion of nondetects at the background well minus the proportion of nondetects at the downgradient well.

```
> ciTableProp(n1 = 36, p1.hat = c(0.2, 0.4, 0.6),
    n2 = 36, p2.hat.minus.p1.hat = c(0.3, 0.15, 0))

                Diff=0.31        Diff=0.14            Diff=0
P1.hat=0.19 [ 0.07, 0.54]  [-0.09, 0.37]  [-0.18, 0.18]
P1.hat=0.39 [ 0.06, 0.55]  [-0.12, 0.39]  [-0.23, 0.23]
P1.hat=0.61 [ 0.09, 0.52]  [-0.10, 0.38]  [-0.23, 0.23]
```

We see that even if the observed difference in the proportion of nondetects is about 15 % points, all of the confidence intervals for the difference between the proportions of nondetects at the two wells contain 0, so if a difference of 15 % points is important to substantiate, we may need to increase our sample sizes.

2.8.3   Nonparametric Confidence Interval for a Percentile

The function `ciNparConfLevel` computes the confidence level associated with a nonparametric confidence interval for the pth quantile (the pth quantile is same as the $100p$th percentile, where $0 \le p \le 1$), given the sample size and value of p. The function `ciNparN` computes the sample size required to achieve a specified confidence level, given the value of p. The function `plotCiNparDesign` plots the relationships between sample size, confidence level, and p.

The data frame `EPA.92c.copper2.df` contains copper concentrations (ppb) at three background wells and two compliance wells.

```
> EPA.92c.copper2.df
```

	Copper.orig	Copper	Censored	Month	Well	Well.type
1	<5	5.0	TRUE	1	1	Background
2	<5	5.0	TRUE	2	1	Background
3	7.5	7.5	FALSE	3	1	Background
...						
38	<5	5.0	TRUE	6	5	Compliance
39	5.6	5.6	FALSE	7	5	Compliance
40	<5	5.0	TRUE	8	5	Compliance

There are eight observations associated with each of the three background wells. Of the 24 observations at the three background wells, 15 are nondetects recorded as "< 5". The other nine observations at the background wells are: 5.4, 5.9, 6.0, 6.1, 6.4, 6.7, 7.5, 8.0, and 9.2. The estimated 95th percentile of copper concentration at the background wells is 7.925 ppb.

```
> Cu.Bkgrd <- with(EPA.92c.copper2.df,
    Copper[Well.type == "Background"]

> eqnpar(Cu.Bkgrd, p = 0.95)

Results of Distribution Parameter Estimation
--------------------------------------------

Assumed Distribution:            None

Estimated Quantile(s):           95'th %ile = 7.925

Quantile Estimation Method:      Nonparametric
```

```
Data:                          Cu.Bkgrd
```

```
Sample Size:                   24
```

If we use the largest observed value of 9.2 as the upper confidence limit of the 95th percentile of the copper concentration, the associated confidence level is 71 %.

```
> ciNparConfLevel(n = 24, p = 0.95, ci.type = "upper")
```

```
[1] 0.708011
```

If only four observations had been taken at each well for a total sample size of $n = 12$, the associated confidence level would have been 46 %.

```
> ciNparConfLevel(n = 12, p = 0.95, ci.type = "upper")
```

```
[1] 0.4596399
```

If we want to construct a nonparametric confidence interval for the 95th percentile of copper concentration with an associated confidence level of at least 95 %, we would need $n = 59$ observations (about 20 observations at each background well).

```
> ciNparN(p = 0.95, ci.type = "upper")
```

```
[1] 59
```

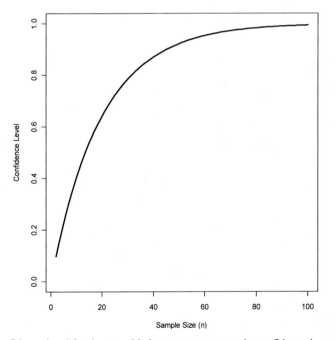

Fig. 2.3 Confidence level for the one-sided upper nonparametric confidence interval for the 95th percentile versus sample size, using the maximum value as the upper confidence limit

Figure 2.3 displays the confidence level of the one-sided upper confidence interval for the 95th percentile as a function of the sample size. To create this plot, type this command:

```
> plotCiNparDesign (p = 0.95, ci.type = "upper",
    range.x.var = c(2, 100), ylim = c(0, 1))
```

2.9   Sample Size for Prediction Intervals

Table 2.2 lists the functions available in **EnvStats** for computing required sample sizes, half-widths, and confidence levels associated with a prediction interval. For the normal distribution, you can compute the half-width of the prediction interval given the user-specified sample size, compute the required sample size given the user-specified half-width, and plot the relationship between sample size and half-width. For a nonparametric prediction interval, you can compute the required sample size for a specified confidence level, compute the confidence level associated with a given sample size, and plot the relationship between sample size and confidence level.

Distribution	Function	Output
Normal	predIntNormHalfWidth	Half-width of prediction interval for normal distribution
	predIntNormN	Required sample size for specified half-width of prediction interval for normal distribution
	plotPredIntNormDesign	Plots for sampling design based on prediction interval for normal distribution
Nonparametric	predIntNparConfLevel predIntNparSimultaneousConfLevel	Confidence level of prediction interval, given sample size
	predIntNparN predIntNparSimultaneousN	Required sample size for specified confidence level of prediction interval
	plotPredIntNparDesign plotPredIntNparSimultaneousDesign	Plots for sampling design based on prediction interval

Table 2.2 Sample size functions for prediction intervals

2.9.1   Prediction Interval for a Normal Distribution

The function `predIntNormHalfWidth` computes the half-width associated with the prediction interval for a normal distribution, given the sample size, number of future observations the prediction interval should contain, estimated standard deviation, and confidence level. The function `predIntNormN` computes the sample size required to achieve a specified half-width, given the number of future observations, estimated standard deviation, and confidence level. The function `plotPredIntNormDesign` plots the relationships between sample size, number of future observations, half-width, estimated standard deviation, and confidence level.

The data frame `EPA.92c.arsenic3.df` contains arsenic concentrations (ppb) collected quarterly for 3 years at a background well and quarterly for 2 years at a compliance well.

```
> EPA.92c.arsenic3.df

    Arsenic Year  Well.type
1      12.6    1 Background
2      30.8    1 Background
...
19      2.6    5 Compliance
20     51.9    5 Compliance
```

The estimated mean and standard deviation for the background well are 28 and 17 ppb, respectively. The exact two-sided 95 % prediction limit for the next $k = 4$ future observations is $[-25, 80]$, which has a half-width of 52.5 ppb and includes values less than 0, which are not possible to observe.

```
> As.Bkgrd <- with(EPA.92c.arsenic3.df,
      Arsenic[Well.type == "Background"])

> predIntNorm(As.Bkgrd, k = 4, method = "exact")

Results of Distribution Parameter Estimation
--------------------------------------------

Assumed Distribution:            Normal

Estimated Parameter(s):          mean = 27.51667
                                 sd   = 17.10119

Estimation Method:               mvue

Data:                            As.Bkgrd

Sample Size:                     12

Prediction Interval Method:      exact

Prediction Interval Type:        two-sided
```

```
Confidence Level:                    95%

Number of Future Observations:    4

Prediction Interval:              LPL = -24.65682
                                  UPL =  79.69015
```

In fact, given an assumed standard deviation of $s = 17$, the smallest half width you can achieve for a prediction interval for the next $k = 4$ future observations is 42 ppb, based on an infinite sample size. ***Unlike a confidence interval, the half-width of a prediction interval does not approach 0 as the sample size increases.*** Figure 2.4 shows a plot of sample size versus half-width for a 95 % prediction interval for a normal distribution for various values of k (the number of future observations), assuming a standard deviation of $s = 17$.

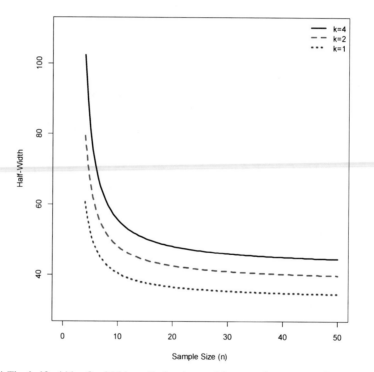

Fig. 2.4 The half-width of a 95 % prediction interval for arsenic concentrations (ppb) as a function of sample size and number of future observations (k), assuming a standard deviation of 17 ppb

Type these commands to create the plot:

```
> plotPredIntNormDesign(range.x.var = c(4, 50), k = 4,
    sigma.hat = 17, xlim = c(0, 50), ylim = c(30, 110),
    main = "")

> plotPredIntNormDesign(range.x.var = c(4, 50), k = 2,
    sigma.hat = 17, plot.col = 2, plot.lty = 2, add = TRUE)

> plotPredIntNormDesign(range.x.var = c(4, 50), k = 1,
    sigma.hat = 17, plot.col = 4, plot.lty = 3, add = TRUE)

> legend("topright", c("k=4", "k=2", "k=1"),
    col = c(1, 2, 4), lty = 1:3, lwd = 3, bty = "n")
```

2.9.2 Nonparametric Prediction Interval

The function `predIntNparConfLevel` computes the confidence level associated with a nonparametric prediction interval, given the minimum number of future observations the interval should contain (k), the number of future observations (m), and sample size. The function `predIntNparN` computes the sample size required to achieve a specified confidence level, given the number of future observations. The function `plotPredIntNparDesign` plots the relationships between sample size, confidence level, and number of future observations.

Table 2.3 shows the required sample size for a two-sided nonparametric prediction interval for the next m future observations (assuming $k = m$) for various values of m and required confidence levels, assuming we are using the minimum and maximum values as the prediction limits. The values for the table are generated using this command:

```
> predIntNparN(m = rep(c(1, 5, 10), 2),
    conf.level = rep(c(0.9, 0.95), each = 3))
```

Confidence level (%)	# future observations (m)	Required sample size (n)
90	1	19
	5	93
	10	186
95	1	39
	5	193
	10	386

Table 2.3 Required sample sizes for a two-sided nonparametric prediction interval, using the minimum and maximum values as the prediction limits

Figure 2.5 displays the confidence level of a two-sided nonparametric prediction interval as a function of sample size for various values of m, using the minimum and maximum values as the prediction limits. To create this figure, type these commands:

```
> plotPredIntNparDesign(range.x.var = c(2, 100), k = 1,
    m = 1, xlim = c(0, 100), ylim = c(0, 1), main = "")

> plotPredIntNparDesign(range.x.var = c(2, 100), k = 5,
    m = 5, plot.col = 2, plot.lty = 2, add = TRUE)

> plotPredIntNparDesign(range.x.var = c(2, 100),
    k = 10, m = 10, plot.col = 4, plot.lty = 3, add = TRUE)

> legend("bottomright", c("m= 1", "m= 5", "m=10"),
    col = c(1, 2, 4), lty = 1:3, lwd = 3)
```

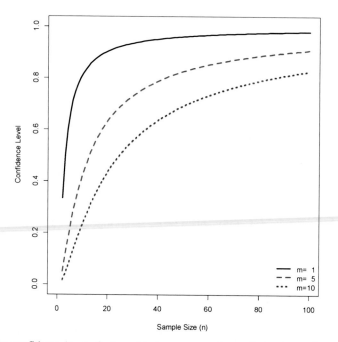

Fig. 2.5 The confidence level of a two-sided nonparametric prediction interval as a function of sample size, for various values of the number of future observations (m)

2.10 Sample Size for Tolerance Intervals

Table 2.4 lists the functions available in EnvStats for computing required sample sizes, half-widths, coverage, and confidence levels associated with a tolerance interval. For the normal distribution, you can compute the half-width of the tolerance interval given the user-specified sample size and coverage, compute the required sample size given the user-specified half-width and coverage, and plot the relationship between sample size, half-width, and coverage. For a non-parametric prediction interval, you can compute the required sample size for a specified confidence level and coverage, compute the confidence level associated with a given sample size and coverage, compute the coverage associated with a given sample size and confidence level, and plot the relationship between sample size, confidence level, and coverage.

Distribution	Function	Output
Normal	tolIntNormHalfWidth	Half-width of tolerance interval for normal distribution
	tolIntNormN	Required sample size for specified half-width of tolerance interval for normal distribution
	plotTolIntNormDesign	Plots for sampling design based on tolerance interval for normal distribution
Nonparametric	tolIntNparConfLevel	Confidence level of tolerance interval, given the coverage and sample size
	tolIntNparCoverage	Coverage of tolerance interval, given confidence level and sample size
	tolIntNparN	Required sample size for specified confidence level and coverage of a tolerance interval
	plotTolIntNparDesign	Plots for sampling design based on a tolerance interval

Table 2.4 Sample size functions for tolerance intervals

2.10.1 Tolerance Interval for a Normal Distribution

The function tolIntNormHalfWidth computes the half-width associated with a tolerance interval for a normal distribution, given the sample size, coverage, estimated standard deviation, and confidence level. The function tolIntNormN computes the sample size required to achieve a specified half-width, given the coverage, estimated standard deviation, and confidence level. The function plotTolIntNormDesign plots the relationships between sample size, half-width, coverage, estimated standard deviation, and confidence level.

Again using the data frame EPA.92c.arsenic3.df containing arsenic concentrations, we saw in Sect. 2.9.1 that the estimated mean and standard deviation for the background well are 28 and 17 ppb, respectively, based on a sample size of $n = 12$ quarterly samples. The two-sided β-content tolerance limit with 95 % coverage and associated confidence level of 99 % is [−39, 94], which has a half-width of 66.5 ppb and includes values less than 0, which are not possible to observe.

```
> tolIntNorm(As.Bkgrd, coverage = 0.95,
    cov.type = "content", conf.level = 0.99)
```

```
Results of Distribution Parameter Estimation
---------------------------------------------

Assumed Distribution:              Normal

Estimated Parameter(s):            mean = 27.51667
                                   sd   = 17.10119

Estimation Method:                 mvue

Data:                              As.Bkgrd

Sample Size:                       12

Tolerance Interval Coverage:       95%

Coverage Type:                     content

Tolerance Interval Method:         Wald-Wolfowitz Approx

Tolerance Interval Type:           two-sided

Confidence Level:                  99%

Tolerance Interval:                LTL = -38.66445
                                   UTL =  93.69778
```

In fact, given an assumed standard deviation of $s = 17$, the smallest half width you can achieve for a tolerance interval with 95 % coverage and 99 % confidence is 33 ppb, based on an infinite sample size. *Unlike a confidence interval, the half-width of a tolerance interval does not approach 0 as the sample size increases.* Figure 2.6 shows a plot of sample size versus half-width for a β-content tolerance interval for a normal distribution with confidence level 99 % for various values of coverage, assuming a standard deviation of $s = 17$. It was created with these commands:

```
> plotTolIntNormDesign(range.x.var = c(5, 50),
    sigma.hat = 17, coverage = 0.99, conf = 0.99,
    xlim = c(0, 50), ylim = c(0, 200), main = "")

> plotTolIntNormDesign(range.x.var = c(5, 50),
    sigma.hat = 17, coverage = 0.95, conf = 0.99,
    plot.col = 2, plot.lty = 2, add = TRUE)
```

```
> plotTolIntNormDesign(range.x.var = c(5, 50),
    sigma.hat = 17, coverage = 0.90, conf = 0.99,
    plot.col = 4, plot.lty = 3, add = TRUE)

> legend("topright",
    paste(c("99%", "95%", "90%"), "Coverage"),
    col = c(1, 2, 4), lty=1:3, lwd=3, bty = "n")
```

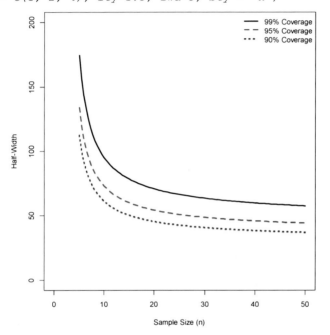

Fig. 2.6 The half-width of a tolerance interval for arsenic concentrations (ppb) as a function of sample size and coverage, assuming a standard deviation of 17 ppb

2.10.2 Nonparametric Tolerance Interval

The function `tolIntNparConfLevel` computes the confidence level associated with a nonparametric tolerance interval, given the coverage and sample size. The function `tolIntNparCoverage` computes the coverage associated with the tolerance interval, given the confidence level and sample size. The function `tolIntNparN` computes the sample size required to achieve a specified confidence level, for a given coverage. The function `plotTolIntNparDesign` plots the relationships between sample size, confidence level, and coverage.

Table 2.5 shows the required sample size for a two-sided nonparametric tolerance interval for various values of coverage and required confidence levels, assuming we are using the minimum and maximum values as the tolerance limits. The values for the table are generated using this command:

```
> tolIntNparN(coverage = rep(c(0.8, 0.9, 0.95), 2),
    conf.level = rep(c(0.9, 0.95), each = 3))
```

Confidence level (%)	Coverage (%)	Required sample size (*n*)
90	80	18
	90	38
	95	77
95	80	22
	90	46
	95	93

Table 2.5 Required sample sizes for a two-sided nonparametric tolerance interval, using the minimum and maximum values as the tolerance limits

Figure 2.7 displays the confidence level of a two-sided nonparametric tolerance interval as a function of sample size for various values of coverage, using the minimum and maximum values as the tolerance limits.

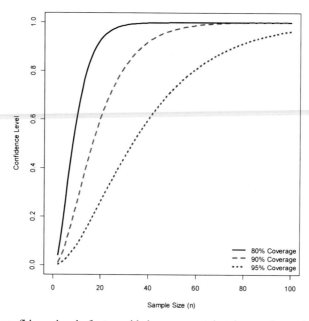

Fig. 2.7 The confidence level of a two-sided nonparametric tolerance interval as a function of sample size, for various values of coverage

To create this plot, type these commands:

```
> plotTolIntNparDesign(range.x.var = c(2, 100),
    coverage = 0.8, xlim = c(0, 100), ylim = c(0, 1),
    main = "")
```

```
> plotTolIntNparDesign(range.x.var = c(2, 100),
    coverage = 0.90, plot.col = 2, plot.lty = 2, add = TRUE)

> plotTolIntNparDesign(range.x.var = c(2, 100),
    coverage = 0.95, plot.col = 4, plot.lty = 3, add = TRUE)

> legend("bottomright",
    paste(c("80%", "90%", "95%"), "Coverage"),
    col = c(1, 2, 4), lty = 1:3, lwd = 3, bty = "n")
```

2.11 Sample Size and Power for Hypothesis Tests

Table 2.6 lists the functions available in EnvStats for computing required sample sizes, powers, and minimal detectable differences associated with several different hypothesis tests. In this section, we will illustrate how to use EnvStats functions to explore the relationship between sample size and power for testing the mean of a normal distribution, testing a binomial proportion, and using simultaneous prediction limits with retesting. See Millard et al. (2014) and the help files for the functions listed in Table 2.6 for more examples of exploring the relationship between sample size and power for other kinds of hypothesis tests.

Test	Function	Output
Student's t-test	tTestPower	Power of t-test
	tTestAlpha	Type I error of t-test
	tTestN	Required sample size for specified power of t-test
	tTestScaledMdd	Required scaled minimal detectable difference (δ/σ) for specified power of t-test
	plotTTestDesign	Plots for sampling design based on t-test
Student's t-test, lognormal distribution	tTestLnormAltPower	Power of one- or two-sample t-test assuming lognormal distribution
	tTestLnormAltN	Required sample size for specified power for one- or two-sample t-test assuming lognormal distribution
	tTestLnormAltRatioOfMeans	Required ratio of means for specified power for one- or two-sample t-test assuming lognormal distribution
	plotTTestLnormAltDesign	Plots for sampling design based on one- or two-sample t-test assuming lognormal distribution

Table 2.6 Sample size and power functions for hypothesis tests

Test	Function	Output
ANOVA F-test	`aovPower`	Power of F-test for one-way ANOVA
	`aovN`	Required sample size for specified power of F-test for one-way ANOVA
	`plotAovDesign`	Plots for sampling design based on F-test for one-way ANOVA
Proportion test, binomial distribution	`propTestPower`	Power of one- or two-sample proportion test
	`propTestN`	Required sample size for specified power for one- or two-sample proportion test
	`propTestMdd`	Required minimal detectable difference for specified power for one- or two-sample proportion test
	`plotPropTestDesign`	Plots for sampling design based on one- or two-sample proportion test
Linear trend t-test	`linearTrendTestPower`	Power of test for non-zero slope
	`linearTrendTestN`	Required sample size for specified power for test of non-zero slope
	`linearTrendTestScaledMds`	Required minimal detectable slope for specified power for test of non-zero slope
	`plotLinearTrendTestDesign`	Plots for sampling design based on test for non-zero slope
Prediction interval, normal distribution	`predIntNormTestPower`	Power of test based on prediction interval for normal distribution
	`plotPredIntNormTestPowerCurve`	Power curve for test based on prediction interval for normal distribution
	`predIntNormSimultaneousTestPower`	Power of test based on simultaneous prediction interval for normal distribution
	`plotPredIntNormSimultaneousTestPowerCurve`	Power curve for test based on simultaneous prediction interval for normal distribution

Table 2.6 (continued). Sample size and power functions for hypothesis tests

Test	Function	Output
Prediction interval, lognormal distribution	`predIntLnormAltTestPower`	Power of test based on prediction interval for lognormal distribution
	`plotPredIntLnormAltTestPowerCurve`	Power curve for test based on prediction interval for lognormal distribution
	`predIntLnormAltSimultaneousTestPower`	Power of test based on simultaneous prediction interval for lognormal distribution
	`plotPredIntLnormAltSimultaneousTest PowerCurve`	Power curve for test based on simultaneous prediction interval for lognormal distribution
Prediction interval, nonparametric	`predIntNparSimultaneousTestPower`	Power of test based on nonparametric simultaneous prediction interval
	`plotPredIntNparSimultaneousTestPower Curve`	Power curve for test based on nonparametric simultaneous prediction interval

Table 2.6 (continued). Sample size and power functions for hypothesis tests

2.11.1 Testing the Mean of a Normal Distribution

Power and sample size calculations based on Student's t-test involve four quantities:

1. The fixed type I error (also called the α-level).
2. The desired power of the test.
3. The sample size(s).
4. The scaled minimal detectable difference (scaled MDD), also often called the effect size. For the one-sample case, the scaled MDD is the difference between the true population mean and the population mean hypothesized under the null hypothesis, divided by the population standard deviation. For the two-sample case, the scaled MDD is the difference between the true population means for the two groups minus the difference between the population means hypothesized for the two groups under the null hypothesis, divided by the population standard deviation (the standard deviation is assumed to be the same for both groups). Because the term "effect size" is sometimes used to denote simply the difference between the means, we always use the term scaled MDD to denote this quantity.

The **EnvStats** function `tTestPower` computes the power associated with the Student's t-test to perform a hypothesis test for the mean of a normal distribution or the difference between two means, given the sample size, scaled MDD, and α-level. The function `tTestAlpha` computes the α-level given the power, sample size, and scaled MDD. The function `tTestN` computes the sample size required to achieve a specified power, given the scaled MDD and α-level. The function `tTestScaledMdd` computes the scaled MDD associated with user-specified values of power, sample size, and α-level. The function `plotTTestDesign` plots the relationships between sample size, power, scaled MDD, and α-level.

The guidance document *Statistical Analysis of Ground-Water Monitoring Data at RCRA Facilities: Unified Guidance* (USEPA 2009) contains an example on pages 22–6 to 22–8 that uses vinyl chloride (ppb) concentrations at two different compliance wells. There are 4 years of quarterly observations at each of the two wells. The first year of data corresponds to the background period and the subsequent 3 years correspond to the compliance period. The data in this example are stored in the data frame `EPA.09.Ex.22.1.VC.df`.

```
> EPA.09.Ex.22.1.VC.df
```

```
   Year Quarter     Period Well VC.ppb
1    1      1      1 Background GW-1   6.3
2    1      1      2 Background GW-1   9.5
...
31   4      4      3 Compliance GW-2   7.5
32   4      4      4 Compliance GW-2   9.7
```

The groundwater protection standard (GWPS) has been set to 5 ppb. During compliance monitoring, we want to test the null hypothesis that the mean vinyl chloride concentration is less than or equal to 5 ppb versus the alternative that it is greater than 5 ppb based on using 1 year of data (i.e., four quarterly observations). We want to have 80 % power of detecting an increase of twice the GWPS (i.e., detecting a true mean vinyl chloride concentration of 10 ppb, a difference of 5 ppb between the assumed mean under the null hypothesis and the mean under the alternative hypothesis).

In this example, first we'll use the first year (background period) of monitoring to estimate the standard deviation of vinyl chloride measurements to determine the required α-level. Then we'll see how changing the α-level and sample size affects the power.

For the first year (background period) of monitoring, the observed means and standard deviations are 8.9 and 2.4 ppb for Well 1, and 7.4 and 3.9 ppb for Well 2, and the pooled estimate of standard deviation (assuming the standard deviation is the same at the two wells) is 3.2 ppb.

```
> summaryStats(VC.ppb ~ Well, data = EPA.09.Ex.22.1.VC.df,
      subset = Period == "Background", digits = 1)
```

2.11. Sample Size and Power for Hypothesis Tests

```
       N Mean   SD Median Min   Max
GW-1 4   8.9 2.4      8.8 6.3 11.9
GW-2 4   7.4 3.9      7.4 3.0 12.0
```

```
> VC.lm.fit <- lm(VC.ppb ~ Well,
    data = EPA.09.Ex.22.1.VC.df,
    subset = Period == "Background")
```

```
> summary(VC.lm.fit)$sigma
```

```
[1] 3.200976
```

However, if we compute a two-sided 95 % confidence interval for the true standard deviation based on the background period data, we see that it may be as high as about 6 ppb:

```
> sqrt(enorm(VC.lm.fit$residuals, ci = TRUE,
    ci.param = "variance")$interval$limits)
```

```
     LCL        UCL
1.959408 6.031586
```

Assuming population standard deviations of 3.2 and 6 ppb, basing the one-sample t-test on $n = 4$ observations, we need to set the type I error level to 0.057 or 0.23 respectively in order to achieve 80 % power of detecting a true concentration of vinyl chloride of 10 ppb:

```
> tTestAlpha(n.or.n1 = 4, delta.over.sigma = 5 / c(3, 6),
    power = 0.8, sample.type = "one.sample",
    alternative = "greater")
```

```
[1] 0.05763283 0.22936065
```

If we set the significance level to 1 % and assume a standard deviation of 3.2 ppb, we can see how the power varies with sample size:

```
> tTestPower(n.or.n1 = c(4, 8, 12),
    delta.over.sigma = 5 / 3.2, alpha = 0.01,
    alternative = "greater")
```

```
[1] 0.3173891 0.8839337 0.9911121
```

If we set the significance level to 1 %, the desired power to 90 %, and assume a standard deviation of 6 ppb, we would need a sample size of at least $n = 22$ to detect an average vinyl chloride concentration that is 5 ppb above the GWPS:

```
> tTestN(delta.over.sigma = 5 / 6, alpha = 0.01,
    power = 0.9, alternative = "greater")
```

```
[1] 22
```

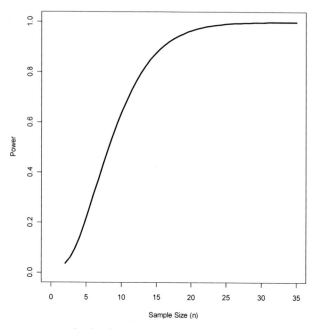

Fig. 2.8 Power versus sample size for a one-sample t-test with a significance level of 1 %, assuming a scaled minimal detectable difference of $\delta/\sigma = 1$

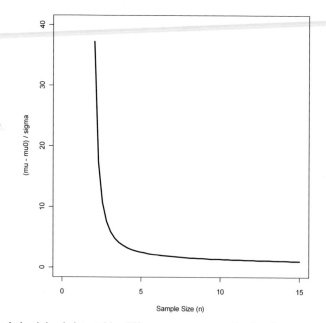

Fig. 2.9 Scaled minimal detectable difference versus sample size for a one-sample t-test with a significance level of 1 %, assuming a power of 90 %

Figure 2.8 plots power as a function of sample size for a significance level of 1 %, assuming a scaled minimal detectable difference of 1. Use this command to produce it:

```
> plotTTestDesign(alpha = 0.01, delta.over.sigma = 1,
    range.x.var = c(2, 35), xlim = c(0, 35), ylim = c(0, 1),
    alternative="greater", approx = FALSE)
```

Figure 2.9 plots the scaled minimal detectable difference as a function of sample size for significance level of 1 %, assuming a power of 90 %.

```
> plotTTestDesign(y.var = "delta.over.sigma",
    alpha = 0.01, power = 0.9, range.x.var = c(2, 15),
    xlim = c(0, 15), ylim = c(0, 40), alternative="greater",
    approx. = FALSE)
```

2.11.2 Testing a Binomial Proportion

The guidance document *Statistical Analysis of Ground-Water Monitoring Data at RCRA Facilities: Unified Guidance* (USEPA 2009) contains an example on page 22–20 that involves determining whether more than 10 % of chlorine gas containers are stored at pressures above a compliance limit. We want to test the one-sided null hypothesis that 10 % or fewer of the containers are stored at pressures greater than the compliance limit versus the alternative that more than 10 % are stored at pressures greater than the compliance limit. We want to have at least 90 % power of detecting a true proportion of 30 % or greater, using a 5 % Type I error level. The example in the guidance document uses the normal approximation to the binomial distribution (without a continuity correction) to determine we need to check 30 containers:

```
> propTestN(p.or.p1 = 0.3, p0.or.p2 = 0.1, alpha = 0.05,
    power = 0.9, sample.type = "one.sample",
    alternative = "greater", approx = TRUE, round.up = TRUE)

[1] 30
```

However, a quick simulation shows that the true Type I error of the hypothesis test based on the normal approximation without using the continuity correction is inflated above 5 % and is really about 7 %:

```
> set.seed(274)

> N <- 10000

> Reject.vec <- logical(N)
```

```
> for(i in 1:N) {
    Reject.vec[i] <- prop.test(
    x = rbinom(n = 1, size = 30, prob = 0.1), n = 30, p = 0.1,
    alternative = "greater", correct = FALSE)$p.value < 0.05
    }
```

```
> mean(Reject.vec)
```

```
[1] 0.071
```

The 95 % confidence interval for the true Type I error level based on our simulation of 10,000 trials is [6.6 %, 7.6 %]:

```
> binom.test(x = sum(Reject.vec), n = length(Reject.vec),
    p = 0.05)$conf.int
```

```
[1] 0.06604181 0.07620974
```

We could try basing our sample size calculation on the test based on the normal approximation **with** the continuity correction, but simulation shows that the continuity correction makes the true Type I error rate about 2.5 % with a 95 % confidence interval of [2.2 %, 2.8 %] for the true Type I error rate:

```
> set.seed(538)
```

```
> N <- 10000
```

```
> Reject.vec <- logical(N)
```

```
> for(i in 1:N) {
    Reject.vec[i] <- prop.test(
    x = rbinom(n = 1, size = 30, prob = 0.1), n = 30, p = 0.1,
    alternative = "greater", correct = TRUE)$p.value < 0.05
    }
```

```
> mean(Reject.vec)
```

```
[1] 0.0248
```

```
> binom.test(x = sum(Reject.vec), n = length(Reject.vec),
    p = 0.05)$conf.int
```

```
[1] 0.02184098 0.02803999
```

If we base our sample size calculation on the exact binomial test instead of the test based on the normal approximation, we can set how much the actual Type I error rate can deviate from what we specify by using the tol.alpha argument to propTestN. By default, tol.alpha is equal to 10 % of the value of alpha, so in this case tol.alpha=0.005 which means the smallest the true Type I error rate can be is 0.045, and the required sample size is 34:

```
> propTestN(p.or.p1 = 0.3, p0.or.p2 = 0.1, alpha = 0.05,
    power = 0.9, sample.type = "one.sample",
    alternative = "greater", approx = FALSE, round.up = TRUE)
$n
[1] 34

$power
[1] 0.9214717

$alpha
[1] 0.04814433

$q.critical.upper
[1] 6
```

If we allow the true Type I error to deviate by 0.01, the required sample size is 33:

```
> propTestN(p.or.p1 = 0.3, p0.or.p2 = 0.1, alpha = 0.05,
    power = 0.9, sample.type = "one.sample", tol.alpha = 0.01,
    alternative = "greater", approx = FALSE, round.up = TRUE)
$n
[1] 33

$power
[1] 0.9055545

$alpha
[1] 0.04170385

$q.critical.upper
[1] 6
```

2.11.3 Testing Multiple Wells for Compliance with Simultaneous Prediction Intervals

The guidance document *Statistical Analysis of Ground-Water Monitoring Data at RCRA Facilities: Unified Guidance* (USEPA 2009) contains an example on page 19–23 that involves monitoring $n_w = 100$ compliance wells at a large facility with minimal natural spatial variation every 6 months for $n_c = 20$ separate chemicals. There are $n = 25$ background measurements for each chemical to use to create simultaneous prediction intervals. We would like to determine which kind of resampling plan based on normal distribution simultaneous prediction intervals to use (1-of-*m*, 1-of-*m* based on means, or Modified California) in order to have adequate power of detecting an increase in chemical concentration at any of the 100 wells while at the same time maintaining a site-wide false positive rate

(SWFPR) of 10 % per year over all 4,000 (100 wells × 20 chemicals × semi-annual sampling) comparisons.

The EnvStats functions for computing power based on simultaneous prediction limits include the argument r that is the number of future sampling occasions (r=2 in this case because we are performing semi-annual sampling), so to compute the individual test Type I error level α_{test} (and thus the individual test confidence level), we only need to worry about the number of wells (100) and the number of constituents (20): $\alpha_{test} = 1-(1-\alpha)^{1/(nw \times nc)}$. The individual confidence level is simply $1-\alpha_{test}$. Plugging in 0.1 for α, 100 for nw, and 20 for nc yields an individual test confidence level of $1-\alpha_{test} = 0.9999473$.

```
> nc <- 20

> nw <- 100

> conf.level <- (1 - 0.1)^(1 / (nc * nw))

> conf.level

[1] 0.9999473
```

Now we can compute the power of any particular sampling strategy using the EnvStats function predIntNormSimultaneousTestPower. For example, to compute the power of detecting an increase of three standard deviations in concentration using the prediction interval based on the "1-of-2" resampling rule, type this command:

```
> predIntNormSimultaneousTestPower(n = 25, k = 1,
    m = 2, r = 2, rule = "k.of.m", delta.over.sigma = 3,
    pi.type = "upper", conf.level = conf.level)

[1] 0.3900202
```

The following commands will reproduce the table shown in Step 2 on page 19–23 of the EPA guidance document:

```
> rule.vec <- c(rep("k.of.m", 3), "Modified.CA",
    rep("k.of.m", 3))

> m.vec <- c(2, 3, 4, 4, 1, 2, 1)

> n.mean.vec <- c(rep(1, 4), 2, 2, 3)

> n.scenarios <- length(rule.vec)

> K.vec <- numeric(n.scenarios)

> Power.vec <- numeric(n.scenarios)

> K.vec <- predIntNormSimultaneousK(n = n, k = 1, m = m.vec,
    n.mean = n.mean.vec, r = r, rule = rule.vec,
    pi.type = "upper", conf.level = conf.level)
```

```
> Power.vec <- predIntNormSimultaneousTestPower(n = n, k = 1,
    m = m.vec, n.mean = n.mean.vec, r = r, rule = rule.vec,
    delta.over.sigma = 3, pi.type = "upper",
    conf.level = conf.level)

> data.frame(Rule = rule.vec, k = rep(1, n.scenarios),
    m = m.vec, N.Mean = n.mean.vec, K = round(K.vec, 2),
    Power = round(Power.vec, 2),
    Total.Samples = m.vec * n.mean.vec)
```

	Rule	k	m	N.Mean	K	Power	Total.Samples
1	k.of.m	1	2	1	3.16	0.39	2
2	k.of.m	1	3	1	2.33	0.65	3
3	k.of.m	1	4	1	1.83	0.81	4
4	Modified.CA	1	4	1	2.57	0.71	4
5	k.of.m	1	1	2	3.62	0.41	2
6	k.of.m	1	2	2	2.33	0.85	4
7	k.of.m	1	1	3	2.99	0.71	3

The above table shows the κ-multipliers for each prediction interval, along with the power of detecting a change in concentration of three standard deviations at any of the 100 wells during the course of a year, for each of the sampling strategies considered. The last three rows of the table correspond to sampling strategies that involve using the mean of two or three observations.

Figure 2.10 shows the power curves for the first four sampling strategies. It was created with these commands:

```
> plotPredIntNormSimultaneousTestPowerCurve(n = 25,
    k = 1, m = 4, r = 2, rule="k.of.m", pi.type = "upper",
    conf.level = conf.level,
    xlab = "SD Units Above Background", main = "")

> plotPredIntNormSimultaneousTestPowerCurve(n = 25,
    k = 1, m = 3, r = 2, rule="k.of.m", pi.type = "upper",
    conf.level = conf.level, add = TRUE, plot.col = 2,
    plot.lty = 2)

> plotPredIntNormSimultaneousTestPowerCurve(n = 25,
    k = 1, m = 2, r = 2, rule="k.of.m", pi.type = "upper",
    conf.level = conf.level, add = TRUE, plot.col = 3,
    plot.lty = 3)

> plotPredIntNormSimultaneousTestPowerCurve(n = 25,
    r = 2, rule="Modified.CA", pi.type = "upper",
    conf.level = conf.level, add = TRUE, plot.col = 4,
    plot.lty = 4)
```

```
> legend(0, 1, c("1-of-4", "Modified CA", "1-of-3",
   "1-of-2"), col = c(1, 4, 2, 3), lty = c(1, 4, 2, 3),
   lwd = 2, bty = "n")
```

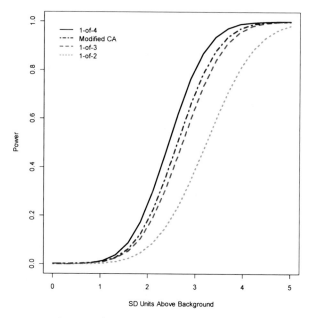

Fig. 2.10 Power versus increase in concentration for various testing strategies based on simultaneous prediction limits, with 100 wells, 20 chemicals, semi-annual sampling, and an annual SWFPR of 10 %

Figure 2.11 shows the power curves for the last three sampling strategies. It was created with these commands:

```
> plotPredIntNormSimultaneousTestPowerCurve(n = 25,
   k = 1, m = 2, n.mean = 2, r = 2, rule="k.of.m",
   pi.type = "upper", conf.level = conf.level,
   xlab = "SD Units Above Background", main = "")
```

```
> plotPredIntNormSimultaneousTestPowerCurve(n = 25,
   k = 1, m = 1, n.mean = 2, r = 2, rule="k.of.m",
   pi.type = "upper", conf.level = conf.level, add = TRUE,
   plot.col = 2, plot.lty = 2)
```

```
> plotPredIntNormSimultaneousTestPowerCurve(n = 25,
   k = 1, m = 1, n.mean = 3, r = 2, rule="k.of.m",
   pi.type = "upper", conf.level = conf.level, add = TRUE,
   plot.col = 3, plot.lty = 3)
```

```
> legend(0, 1, c("1-of-2, Order 2", "1-of-1, Order 3",
   "1-of-1, Order 2"), col = c(1, 3, 2), lty = c(1, 3, 2),
   lwd = 2, bty="n")
```

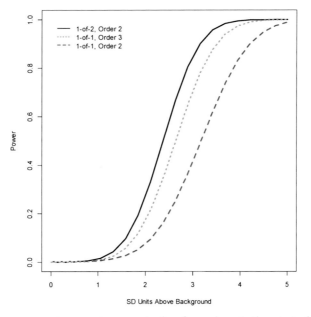

Fig. 2.11 Power versus increase in concentration for various testing strategies based on simultaneous prediction limits using the mean of two measures, with 100 wells, 20 chemicals, semi-annual sampling, and an annual SWFPR of 10 %

2.12 Summary

- The first and most important step of any environmental study is to design the sampling program.
- **Probability sampling** or **random sampling** involves using a random mechanism to select samples from the population. All statistical methods used to quantify uncertainty assume some form of random sampling was used to obtain the sample.
- The Data Quality Objectives (DQO) process is a systematic planning tool based on the scientific method. The last two steps involve trading off limits on Type I and Type II errors and sample size.
- You can use the EnvStats functions listed in Tables 2.1, 2.2, 2.4, and 2.6 (and in the help file *Power and Sample Size*) to estimate required samples sizes for an environmental study.

Chapter 3

Looking at Data

3.1 Introduction

Once you have a collection of observations from your environmental study, you should thoroughly examine the data in as many ways as possible and relevant. When the first widely available commercial statistical software packages came out in the 1960s, the emphasis was on statistical summaries of data, such as means, standard deviations, and measures of skew and kurtosis. It is still true that "a picture is worth a thousand words," and no amount of summary or descriptive statistics can replace a good graph to explain your data. John Tukey coined the acronym **EDA**, which stands for ***Exploratory Data Analysis***. Helsel and Hirsch (1992), USEPA (2006a), and Millard et al. (2014) give a good overview of statistical and graphical methods for exploring environmental data. Cleveland (1993, 1994) and Chambers et al. (1983) are excellent general references for methods of graphing data. This chapter discusses the functions available in ENVSTATS for producing summary statistics and graphs to describe and look at environmental data.

Statistic or plot	Function	Output
Summary statistics	summaryFull	Summary statistics
	summaryStats	Summary statistics, p-values, and confidence intervals
Strip chart	stripChart	Strip chart with confidence intervals for mean or pseudo-median
Probability distribution function (PDF) plot	epdfPlot	Empirical PDF plot
Cumulative distribution function (CDF) plots	ecdfPlot	Empirical CDF plot
	cdfCompare	Compare empirical CDF to a hypothesized CDF, or compare two empirical CDFs
Quantile-Quantile (Q-Q) plots	qqPlot	Q-Q plot comparing data to a theoretical distribution or comparing two data sets
	qqPlotGestalt	Numerous Q-Q plots based on a specified distribution
Box-Cox transformations	boxcox	Determine optimal Box-Cox transformation

Table 3.1 Functions in ENVSTATS for exploratory data analysis

S.P. Millard, *EnvStats: An R Package for Environmental Statistics*,
DOI 10.1007/978-1-4614-8456-1_3, © Springer Science+Business Media New York 2013

3.2   EDA Using ENVSTATS

R comes with numerous functions for producing summary statistics and graphs to look at your data. Table 3.1 lists the additional functions available in ENVSTATS for performing EDA.

Statistic	What it measures / how it is computed	Robust to extreme values?
Mean	Center of distribution Sum of observations divided by sample size Where the histogram balances	No
Trimmed mean	Center of distribution Trim off extreme observations and compute mean Where the trimmed histogram balances	Somewhat
Median	Center of the distribution Middle value or mean of middle values Half of observations are less and half are greater	Very
Geometric mean	Center of distribution Exponentiated mean of log-transformed observations Estimates true median for a lognormal distribution	Yes
Variance	Spread of distribution Average of squared distances from the mean	No
Standard deviation	Spread of distribution Square root of variance In same units as original observations	No
Range	Spread of distribution Maximum minus minimum	No
Interquartile range	Spread of distribution 75th percentile minus 25th percentile Range of middle 50 % of data	Yes
Median absolute deviation	Spread of distribution $1.4826 \times$ Median of distances from the median	Yes
Geometric standard deviation	Spread of distribution Exponentiated standard deviation of log-transformed observations	No
Coefficient of variation	Spread of distribution/center of distribution Standard deviation divided by mean Sometimes multiplied by 100 and expressed as a percentage	No
Skew	How the distribution leans (left, right, or centered) Average of cubed distances from the mean	No
Kurtosis	Peakedness of the distribution Average of quartic distances from the mean, then subtract 3	No

Table 3.2 A description of commonly used summary statistics

3.3   Summary Statistics

Summary statistics (also called *descriptive statistics*) are numbers that you can use to summarize the information contained in a collection of observations. Summary statistics are also called *sample statistics* because they are statistics computed from a sample; they do not describe the whole population.

One way to classify summary or descriptive statistics is by what they measure: location (central tendency), spread (variability), skew (long-tail in one direction), kurtosis (peakedness), etc. Another way to classify summary statistics is by how they behave when unusually extreme observations are present: sensitive versus robust. Table 3.2 summarizes several kinds of descriptive statistics based on these two classification schemes.

Millard et al. (2014) describe functions in R for computing summary statistics. The EnvStats help topic *Summary Statistics* lists additional and/or modified functions for computing summary statistics.

3.3.1   *Summary Statistics for TcCB Concentrations*

The guidance document USEPA (1994b) contains measures of 1,2,3,4-Tetrachlorobenzene (TcCB) concentrations (ppb) from soil samples at a "Reference" site and a "Cleanup" area. The Cleanup area was previously contaminated and we are interested in determining whether the cleanup process has brought the level of TcCB back down to what you would find in soil typical of that particular geographic region. In EnvStats, these data are stored in the data frame `EPA.94b.tccb.df`.

```
> EPA.94b.tccb.df

     TcCB.orig   TcCB Censored      Area
1         0.22   0.22    FALSE Reference
2         0.23   0.23    FALSE Reference
...
47        1.33   1.33    FALSE Reference
48       <0.09   0.09     TRUE   Cleanup
...
123      51.97  51.97    FALSE   Cleanup
124     168.64 168.64    FALSE   Cleanup
```

There are 47 observations from the Reference site and 77 in the Cleanup area. There is one observation in the Cleanup area that was coded as "ND," which stands for nondetect. This means that the concentration of TcCB for this soil sample (if any was present at all) was so small that the procedure used to quantify TcCB concentrations could not reliably measure the true concentration. For the purpose of creating the data frame `EPA.94b.tccb.df`, we set the (unreported) detection limit to the value of the smallest observation, which is 0.09. The column `TcCB.orig` displays how the data were originally recorded, the column `TcCB` contains the original observations, except that the nondetect value is set to the censoring level 0.09, the column `Censored` indicates whether the observation was censored (i.e., reported as a nondetect), and the column `Area` indicates which area the observation comes from.

The summary statistics for the TcCB data are shown in Sect. 1.11.2 of Chap. 1. The summary statistics indicate that the observations for the Cleanup area are extremely skewed to the right: the medians for the two areas are about the same, but the mean for the Cleanup area is much larger, indicating a few or more "outlying" observations. This may be indicative of residual contamination that was missed during the cleanup process. Figures 1.1, 1.2, and 1.3 in Sect. 1.11.3 display the strip charts, histograms and boxplots for the log-transformed TcCB data.

For the remainder of this chapter, we will assume that you have attached the data frame `EPA.94b.tccb.df` to your search list with the command:

```
> attach(EPA.94b.tccb.df)
```

3.4   Strip Charts

The R function `stripchart` creates one-dimensional scatterplots (also called strip plots or strip charts). The EnvStats function `stripChart` is a modification of `stripchart` that allows you to add confidence intervals for the mean or pseudo-median of each group and also display the results of a hypothesis test that the group means are all equal (confidence intervals are discussed in Chap. 5 and hypothesis tests in Chap. 7). Figure 1.1 in Sect. 1.11.3 displays the strip charts for the log-transformed TcCB data by area and includes confidence intervals for the mean TcCB concentration in each area.

3.5   Empirical PDF Plots

Figures 1.2 and 1.3 in Sect. 1.11.3 show histograms and boxplots for the TcCB data. Strip charts, histograms, and boxplots are all graphical tools used to give you an idea of the shape of the underlying probability density function (pdf; see Chap. 4). Another graphical tool for this purpose is an ***empirical pdf plot*** (also called a ***density plot***), and you can use the EnvStats function `epdfPlot` to create these. When a distribution is discrete and can only take on a finite number of values, the empirical pdf plot is the same as the standard relative frequency histogram; that is, each bar of the histogram represents the proportion of the sample equal to that particular number (or category). When a distribution is continuous, the function `epdfPlot` calls the R function `density` to compute the estimated probability density at a number of evenly spaced points between the minimum and maximum values. Figure 3.1 shows the empirical pdf plot for the log-transformed Reference area TcCB data superimposed on a relative frequency histogram. It was created with these commands:

```
> log.TcCB <- log(TcCB[Area == "Reference"])
```

```
> hist(log.TcCB, freq = FALSE, xlim = c(-2, 1),
    col = "grey", xlab = "log [ TcCB (ppb) ]",
    ylab = "Relative Frequency", main = "")

> epdfPlot(log.TcCB, epdf.col = "blue", add = TRUE)
```

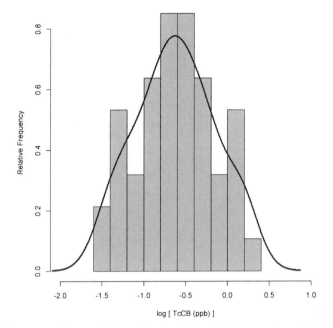

Fig. 3.1 Histogram with overlaid empirical pdf plot for log-transformed Reference area TcCB data

3.6   Quantile (Empirical CDF) Plots

Loosely speaking, the pth **quantile** of a population is the (a) number such that a fraction p of the population is less than or equal to this number. The pth quantile is the same as the $100p$th percentile; for example, the 0.5 quantile is the same as the 50th percentile. For a population, a plot of the quantiles on the x-axis versus the percentage or fraction of the population less than or equal to that number on the y-axis is called a **cumulative distribution function plot** or **cdf plot** (we will talk more about cumulative distribution functions in Chap. 4). The y-axis is usually labeled as the **cumulative probability** or **cumulative frequency**.

When we have a sample of data from some population, we usually do not know what percentiles our observations correspond to because we do not know the true population percentiles, so we use the sample data to estimate them. A **quantile plot** (also called an **empirical cumulative distribution function plot** or **empirical cdf plot**) plots the ordered data (sorted from smallest to largest) on the x-axis versus the estimated cumulative probabilities on the y-axis (Chambers et al. 1983; Cleveland 1993. 1994; Helsel and Hirsch 1992). (Sometimes the x- and y-axes are reversed.) The specific formulas that are used to estimate the cumulative probabilities (also called the ***plotting positions***) are discussed in Millard et al. (2014).

3.6.1   Empirical CDFs for the TcCB Data

Figure 1.4 in Sect. 1.11.4 shows the quantile plot for the Reference area TcCB data. Based on this plot, you can easily pick out the median as about 0.55 ppb and the quartiles as about 0.4 and 0.75 ppb (compare these numbers to the ones listed in Sect. 1.11.2). You can also see that the quantile plot quickly rises, then pretty much levels off after about 0.8 ppb, which indicates that the data are skewed to the right (see the histogram for the Reference area data in Fig. 1.2 in Sect. 1.11.3). Helsel and Hirsch (1992) note that quantile plots, unlike histograms, do not require you to figure out how to divide the data into classes, and, unlike boxplots, all of the data are displayed in the graph.

Figure 1.5 in Sect. 1.11.4 shows the quantile plot for the Reference area TcCB data with a fitted lognormal distribution. We see that the lognormal distribution appears to fit these data quite well.

Figure 1.6 compares the empirical cdf for the Reference area with the empirical cdf for the Cleanup area for the log-transformed TcCB data. As we saw with the histograms and boxplots, the Cleanup area has quite a few extreme values compared to the reference area.

3.7   Probability Plots or Quantile-Quantile (Q-Q) Plots

A ***probability plot*** or ***quantile-quantile (Q-Q) plot*** is a graphical display invented by Wilk and Gnanadesikan (1968) to compare a data set to a particular probability distribution or to compare it to another data set. The idea is that if two population distributions are exactly the same, then they have the same quantiles (percentiles), so a plot of the quantiles for the first distribution versus the quantiles for the second distribution will fall on the 0–1 line (i.e., the straight line $y = x$ with intercept 0 and slope 1). If the two distributions have the same shape and spread but different locations, then the plot of the quantiles will fall on the line $y = a + x$ (parallel to the 0–1 line) where a denotes the difference in locations. If the distributions have different locations and differ by a multiplicative constant b, then the plot of the quantiles will fall on the line $y = a + bx$ (D'Agostino 1986a; Helsel and Hirsch 1992). Various kinds of differences between distributions will yield various kinds of deviations from a straight line. In ENVSTATS, you can add a fitted regression line, a robust regression line, or a 0–1 line to the Q-Q plot.

Instead of adding a fitted regression line to a Q-Q plot, another way to assess deviation from linearity is to use a Tukey mean-difference Q-Q plot, also called an m-d plot (Cleveland 1993). This is a plot of the differences between the quantiles on the y-axis versus the average of the quantiles on the x-axis. If the two sets of quantiles come from the same parent distribution, then the points in an m-d plot should fall roughly along the horizontal line $y = 0$. If one set of quantiles come from the same distribution with a shift in median, then the points in this plot should fall along a horizontal line above or below the line $y = 0$. If the parent distributions of the quantiles differ in scale, then the points on this plot will fall at an angle.

3.7.1   Q-Q Plots for the Normal and Lognormal Distribution

Figure 3.2 shows the normal Q-Q plot for the Reference area TcCB data, along with a fitted regression line. In this figure you can see that the points do not tend to fall on the line, but rather seem to make a U shape. This indicates that the Reference area data are skewed to the right relative to a symmetrical, bell-shaped normal distribution.

```
> qqPlot(TcCB[Area == "Reference"], add.line = TRUE,
    points.col = "blue", ylab = "Quantiles of TcCB (ppb)",
    main = "")
```

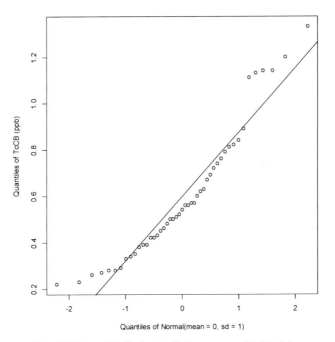

Fig. 3.2 Normal Q-Q plot for Reference area TcCB data

Figure 1.7 in Sect. 1.11.5 shows the normal Q-Q plot for the log-transformed Reference area data, and Fig. 1.8 displays the corresponding Tukey mean-difference Q-Q plot. Here you can see the points do tend to fall on the line, indicating that a lognormal distribution may be a good model for these data. Compare these figures to Fig. 1.5.

An interesting feature of normal Q-Q plots is that if a sample of data comes from a normal distribution, and the data are plotted against quantiles of a standard normal distribution (with mean 0 and variance 1), then the intercept of the fitted line estimates the mean of the population, and the slope of the fitted line estimates the standard deviation (Nelson 1982; Cleveland 1993). For the fitted line in Fig. 1.7, we can eyeball the intercept at about −0.6 and the slope at about 0.5. The actual mean and standard deviation are −0.62 and 0.47, respectively.

3.7.2   Q-Q Plots for Other Distributions

Although they are not as commonly used as Q-Q plots for the normal and lognormal distributions, you can easily create Q-Q plots for other distributions as well. Figure 1.9 in Sect. 1.11.5 shows a gamma Q-Q plot for the reference area TcCB data. As in Fig. 1.7, the points tend to fall on the line, indicating that a gamma distribution may be a good model for these data as well.

As another example, the guidance document *Statistical Analysis of Ground-Water Monitoring Data at RCRA Facilities: Addendum to Interim Final Guidance* (USEPA 1992c) contains a data set of benzene concentrations (ppb) from water samples collected over 6 months from six different background monitoring wells. These data are stored in the data frame EPA.92c.benzene1.df.

```
> EPA.92c.benzene1.df
```

	Benzene.orig	Benzene	Censored	Month	Well
1	<2	2	T	1	1
2	<2	2	T	2	1
...					
35	10	10	F	5	6
36	<2	2	T	6	6

Out of the 36 observations, 33 are reported as "<2", and the other three observations are 10, 12, and 15. The example in the guidance document proposes to model these data as having come from a Poisson distribution, and sets each nondetect to 1 ppb (half the detection limit). (Note: this guidance document has been superseded by USEPA (2009), but we include this example here for illustrative purposes.) Figures 3.3 and 3.4 show the Poisson Q-Q plots for these data, which indicate that the assumption of a Poisson distribution is questionable: there are too many observations with the value 1 (the nondetects), and the detected observations are too large. In Fig. 3.3 we indicate multiple observations that have the same (x, y) coordinates with the number of observations that have those coordinates. In Fig. 3.4 we instead jitter all of the points.

To create the Poisson Q-Q plot shown in Fig. 3.3, type these commands:

```
> attach(EPA.92c.benzene1.df)

> Benzene[Censored] <- 1
```

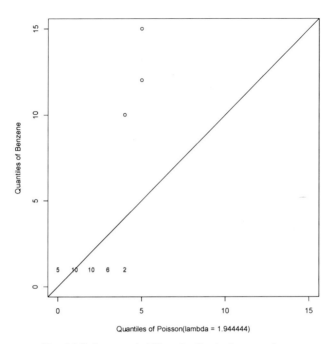

Fig. 3.3 Poisson probability plot for the benzene data

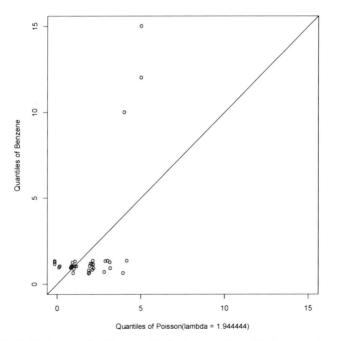

Fig. 3.4 Poisson probability plot for the benzene data with jittered points

```
> qqPlot(Benzene, dist = "pois", estimate.params = TRUE,
    duplicate.points.method = "number", add.line = TRUE,
    qq.line.type = "0-1", points.col = "blue", main = "")
```

To create the Poisson Q-Q plot shown in Fig. 3.4, type these commands:

```
> set.seed(721)
```

```
> qqPlot(Benzene, dist = "pois", estimate.params = TRUE,
    duplicate.points.method = "jitter", add.line = TRUE,
    qq.line.type = "0-1", points.col = "blue", main = "")
```

```
> detach("EPA.92c.benzene1.df")
```

3.7.3   Using Q-Q Plots to Compare Two Data Sets

Besides using Q-Q plots or probability plots to assess whether a set of data appear to come from a particular probability distribution, you can use a Q-Q plot to assess whether two sets of data appear to have the same parent distribution (i.e., the same shape but not necessarily the same location or scale). If the distributions have the same shape (but not necessarily the same location or scale parameters), then the plot will fall roughly on a straight line. If the distributions are exactly the same, then the plot will fall roughly on the straight line $y = x$.

Figure 3.5 shows the Q-Q plot comparing the Cleanup and Reference areas for the log-transformed TcCB data, along with the 0–1 line. It was created with these commands:

```
> qqPlot(log(TcCB[Area == "Reference"]),
    log(TcCB[Area == "Cleanup"]), plot.pos.con = 0.375,
    equal.axes = TRUE, add.line = TRUE,
    qq.line.type = "0-1", points.col = "blue",
    xlab = paste("Quantiles of log [ TcCB (ppb) ]",
      "for Reference Area"),
    ylab = paste("Quantiles of log [ TcCB (ppb) ]",
      "for Cleanup Area"), main = "")
```

In this figure you can see the points do not tend to fall on the 0–1 line, but instead tend to fall along two different lines, both with a steeper slope than 1. Q-Q plots that exhibit this kind of pattern indicate that one of the samples (the Cleanup area data in this case) probably comes from a "mixture" distribution: some of the observations come from a distribution similar in shape and scale to the Reference area distribution, and some of the observations come from a distribution that is shifted to the right and more spread out relative to the Reference area distribution because of residual contamination.

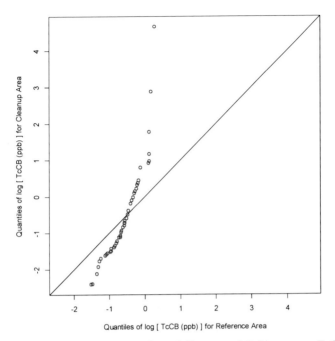

Fig. 3.5 Q-Q plot comparing log-transformed Cleanup and Reference area TcCB data

3.7.4   Building an Internal Gestalt for Q-Q Plots

Probability plots or Q-Q plots are a graphical, subjective way of assessing the goodness-of-fit of a data set to a specified theoretical distribution. In order to be able to assess the goodness-of-fit, you need to have an internal baseline image (a gestalt) of what a "typical" Q-Q plot looks like when in fact the data come from the specified distribution. You can use ENVSTATS to produce numerous Q-Q plots to build up such an internal gestalt (see the help file for qqPlotGestalt for more information).

Figure 3.6 shows a set of four typical normal Q-Q plots based on a sample size of $n = 10$. It was created with these commands:

```
> set.seed(426)

> qqPlotGestalt(num.pages = 1, add.line = TRUE,
    points.col = "blue")
```

Note that with such a small number of observations, there can be a bit of spread about the fitted regression line. Figure 3.7 shows a set of four typical Tukey mean-difference Q-Q plots based on a sample size of $n = 10$.

```
> qqPlotGestalt(num.pages = 1, add.line = TRUE,
    plot.type = "Tukey", estimate.params = TRUE,
    points.col = "blue")
```

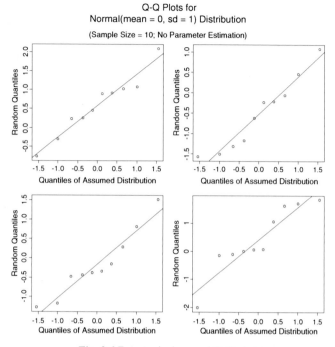

Fig. 3.6 Four typical normal Q-Q plots

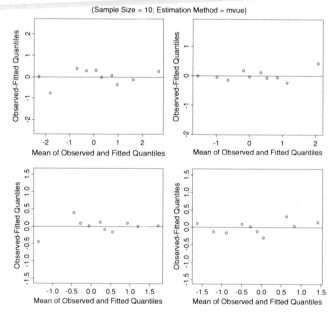

Fig. 3.7 Four typical Tukey mean-difference normal Q-Q plots

3.8   Box-Cox Data Transformations and Q-Q Plots

Two common assumptions for several standard parametric hypothesis tests are:

1.   The observations come from a normal distribution.
2.   If several groups are involved, the variances are the same among all of the groups.

A standard linear regression model makes the above assumptions, and also assumes a linear relationship between the response variable and the predictor variable or variables.

Often, especially with environmental data, the above assumptions do not hold because the original data are skewed and/or they follow a distribution that is not really shaped like a normal distribution. It is sometimes possible, however, to transform the original data so that the transformed observations in fact come from a normal distribution or close to a normal distribution. The transformation may also induce homogeneity of variance and a linear relationship between the response and predictor variable(s) (if this is relevant).

Sometimes, theoretical considerations indicate an appropriate transformation. For example, count data often follow a Poisson distribution, and it can be shown that taking the square root of observations from a Poisson distribution tends to make these data look more bell-shaped (Johnson et al. 1992; Johnson and Wichern 2007; Zar 2010). A common example in the environmental field is that chemical concentration data often appear to come from a lognormal distribution or some other positively skewed distribution. In this case, taking the logarithm of the observations often appears to yield normally distributed data. Usually, a data transformation is chosen based on knowledge of the process generating the data, as well as graphical tools such as quantile-quantile plots and histograms.

Although data analysts knew about using data transformations for several years, Box and Cox (1964) presented a formalized method for deciding on a data transformation. Given a random variable X from some distribution with only positive values, the Box-Cox family of power transformations is defined as:

$$Y = \begin{cases} \dfrac{\left(X^{\lambda} - 1 \right)}{\lambda} , & \lambda \neq 0 \\[2em] \log(X) , & \lambda = 1 \end{cases} \qquad (3.1)$$

where λ (lambda) denotes the power of the transformation and Y is assumed to come from a normal distribution. This transformation is continuous in λ. Note that this transformation also preserves ordering. That is, if $X_1 < X_2$ then $Y_1 < Y_2$.

Box and Cox (1964) proposed choosing the appropriate value of λ based on maximizing the likelihood function. Note that for non-zero values of λ, instead of using the formula of Box and Cox in Eq. 3.1, you may simply use the power trans-formation

$$Y = X^\lambda \tag{3.2}$$

since these two equations differ only by a scale difference and origin shift, and the essential character of the transformed distribution remains unchanged (Draper and Smith 1998).

The value $\lambda = 1$ corresponds to no transformation. Values of λ less than 1 shrink large values of X, and are therefore useful for transforming positively skewed (right-skewed) data. Values of λ larger than 1 inflate large values of X, and are therefore useful for transforming negatively skewed (left-skewed) data (Helsel and Hirsch 1992; Johnson and Wichern 2007). Commonly used values of λ include 0 (log transformation), 0.5 (square-root transformation), −1 (reciprocal), and −0.5 (reciprocal root).

Transformations are not "tricks" used by the data analyst to hide what is going on, but rather useful tools for understanding and dealing with data (Berthouex and Brown 2002). Hoaglin (1988) discusses "hidden" transformations that are used every day, such as the pH scale for measuring acidity. It is often recommend that when dealing with several similar data sets, it is best to find a common transfor-mation that works reasonably well for all the data sets, rather than using slightly different transformations for each data set (Helsel and Hirsch 1992; Shumway et al. 1989).

One problem with data transformations is that translating results on the trans-formed scale back to the original scale is not always straightforward. Estimating quantities such as means, variances, and confidence limits in the transformed scale and then transforming them back to the original scale usually leads to biased and inconsistent estimates (Gilbert 1987; van Belle et al. 2004). For example, exponentiating the confidence limits for a mean based on log-transformed data does not yield a confidence interval for the mean on the original scale. Instead, this yields a confidence interval for the median. It should be noted, however, that quantiles (percentiles) and rank-based procedures are invariant to monotonic transformations (Helsel and Hirsch 1992).

You can use ENVSTATS to determine an "optimal" Box-Cox transformation, based on one of three possible criteria:

- Probability Plot Correlation Coefficient (PPCC)
- Shapiro-Wilk Goodness-of-Fit Test Statistic (W)
- Log-Likelihood Function

You can also compute the value of the selected criterion for a range of values of the transform power λ.

Figure 3.8 displays a plot of the probability plot correlation coefficient versus various values of the transform power λ for the Reference area TcCB data. For this data set, the PPCC reaches its maximum at about $\lambda = 0$, which corresponds to a log transformation. Besides plotting the objective function versus λ, you can also generate Q-Q plots and Tukey mean-difference Q-Q plots for each of the values of λ.

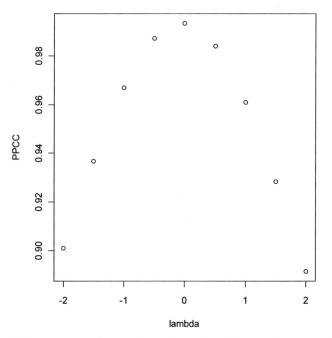

Fig. 3.8 Probability plot correlation coefficient versus Box-Cox transform power (λ) for the Reference area TcCB data

To create Figure 3.8, type these commands:

```
> boxcox.list <- boxcox(TcCB[Area == "Reference"])
> plot(boxcox.list, main = "")
```

To produce the nine Q-Q plots associated with each of the values of λ, type:

```
> plot(boxcox.list, plot.type = "Q-Q")
```

and to produce the nine Tukey mean-difference Q-Q plots type:

```
> plot(boxcox.list, plot.type = "Tukey")
```

To create all of the plots, you can type:

```
> plot(boxcox.list, plot.type = "All")
```

To print the results, type

```
> boxcox.list
```

```
Results of Box-Cox Transformation
---------------------------------

Objective Name:                          PPCC

Data:                                    TcCB[Area == "Reference"]

Sample Size:                             47

   lambda        PPCC
    -2.0  0.9008498
    -1.5  0.9366847
    -1.0  0.9669707
    -0.5  0.9871430
     0.0  0.9932857
     0.5  0.9839249
     1.0  0.9608911
     1.5  0.9284576
     2.0  0.8914832
```

3.9 Summary

- Summary or descriptive statistics can be classified by what they measure (location, spread, skew, kurtosis, etc.) and also how they behave when unusually extreme observations are present (sensitive versus robust).
- Because environmental data usually involve measures of chemical concentrations, and concentrations cannot fall below 0, environmental data often tend to be positively skewed.
- Graphical displays are usually far superior to summary statistics for conveying information in a data set.
- For conveying the distribution of univariate data, use strip plots, histograms, density plots (also called empirical probability distribution function plots), boxplots, and quantile plots (also called empirical cumulative distribution function plots).
- To compare two data sets or to compare a data set to a theoretical probability distribution, use quantile-quantile (Q-Q) plots (also called probability plots), and Tukey mean-difference Q-Q plots.
- You can use Box-Cox transformations along with Q-Q plots to determine a transformation that may satisfy the assumption of normality if this assumption is necessary for a hypothesis test or confidence interval.
- You can use the ENVSTATS functions listed in Table 3.1 to create summary statistics, strip charts with confidence intervals, empirical pdf plots, empirical cdf plots, Q-Q plots, and determine "optimal" Box-Cox transformations.

Chapter 4

Probability Distributions

4.1 Introduction

As we stated in Chap. 2, a population is defined as the entire collection of measurements about which we want to make a statement, such as all possible measurements of dissolved oxygen in a specific section of a stream within a certain time period. Probability distributions are idealized mathematical models that are used to model the variability inherent in a population. Certain probability distributions come up again and again in environmental statistics. This chapter discusses the functions available in ENVSTATS for plotting probability distributions, computing quantities associated with these distributions, and generating random numbers from these distributions. See Millard et al. (2014) for a more in-depth discussion of probability distributions.

Table 4.1 lists the probability distributions available in R and ENVSTATS (see the ENVSTATS help file *Probability Distributions*) and Fig. 4.1 displays examples of the probability density functions for these probability distributions. Most of these distributions are already available in R, but many have been added in ENVSTATS. The help file for `Distribution.df` contains more extensive tables that include the distribution name, abbreviation, type (continuous, discrete, finite discrete, mixed), range (i.e., support), parameters, default values for the parameters, parameter ranges, and estimation methods available for the parameters.

Distribution name	Abbreviation	Parameter(s)
Beta	`beta`	`shape1, shape2, ncp`
Binomial	`binom`	`size, prob`
Cauchy	`cauchy`	`location, scale`
Chi*	`chi`	`df`
Chi-square	`chisq`	`df, ncp`
Empirical*	`emp`	
Exponential	`exp`	`rate`
Extreme value*	`evd`	`location, scale`
F	`f`	`df1, df2, ncp`
Gamma	`gamma`	`shape, scale or rate`
Gamma (alternative parameterization)*	`gammaAlt`	`mean, cv`

Table 4.1 Distribution abbreviations and parameters (*part of EnvStats)

S.P. Millard, *EnvStats: An R Package for Environmental Statistics*,
DOI 10.1007/978-1-4614-8456-1_4, © Springer Science+Business Media New York 2013

Distribution name	Abbreviation	Parameter(s)
Generalized extreme value*	gevd	location, scale, shape
Geometric	geom	prob
Hypergeometric	hyper	m, n, k
Logistic	logis	location, scale
Lognormal	lnorm	meanlog, sdlog
Lognormal (alternative parameterization)*	lnormAlt	mean, cv
Lognormal mixture*	lnormMix	meanlog1, sdlog1, meanlog2, sdlog2, p.mix
Lognormal mixture (alternative parameterization)*	lnormMixAlt	mean1, sd1, mean2, sd2, cv
3-Parameter lognormal*	lnorm3	meanlog, sdlog, threshold
Truncated lognormal*	lnormTrunc	meanlog, sdlog, min, max
Truncated lognormal (alternative parameterization)*	lnormAltTrunc	mean, cv, min, max
Negative binomial	nbinom	size, prob
Normal	norm	mean, sd
Normal mixture*	normMix	mean1, sd1, mean2, sd2, p.mix
Truncated normal*	normTrunc	mean, sd, min, max
Pareto*	pareto	location, shape
Poisson	pois	lambda
Student's t	t	df, ncp
Triangular*	tri	min, max, mode
Uniform	unif	min, max
Weibull	weibull	shape, scale
Wilcoxon rank sum	wilcox	m, n
Zero-modified lognormal (delta)*	zmlnorm	meanlog, sdlog, p.zero
Zero-Modified Lognormal (delta) (alternative parameterization)*	zmlnormAlt	mean, cv, p.zero
Zero-modified normal*	zmnorm	mean, sd, p.zero

Table 4.1 (continued). Distribution abbreviations and parameters (*part of EnvStats)

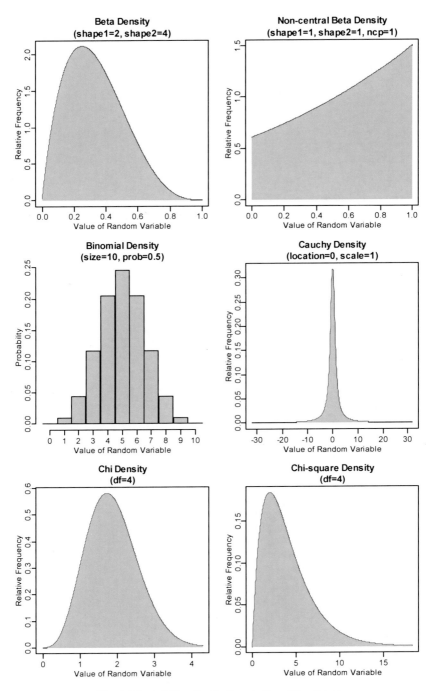

Fig. 4.1 Probability distributions in R and ENVSTATS

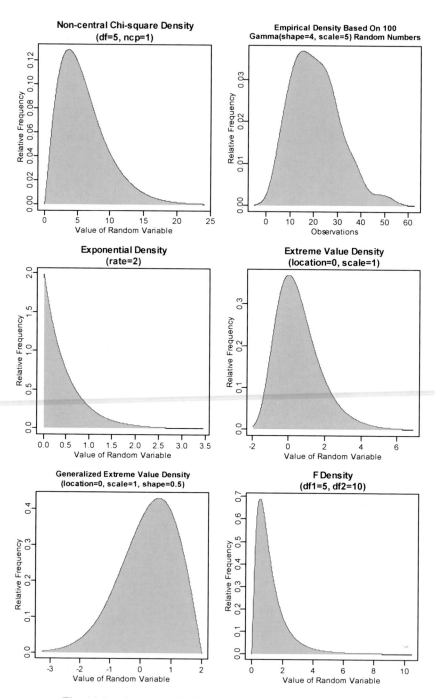

Fig. 4.1 (continued). Probability distributions in R and ENVSTATS

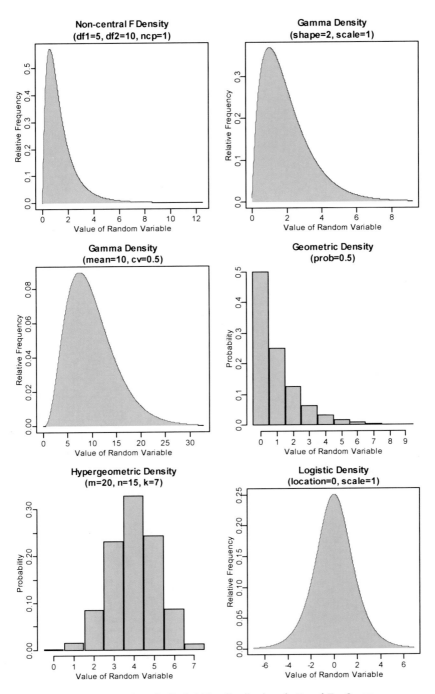

Fig. 4.1 (continued). Probability distributions in R and ENVSTATS

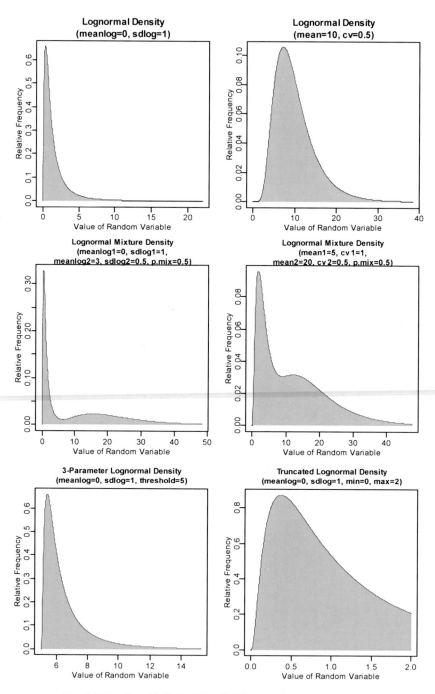

Fig. 4.1 (continued). Probability distributions in R and EnvStats

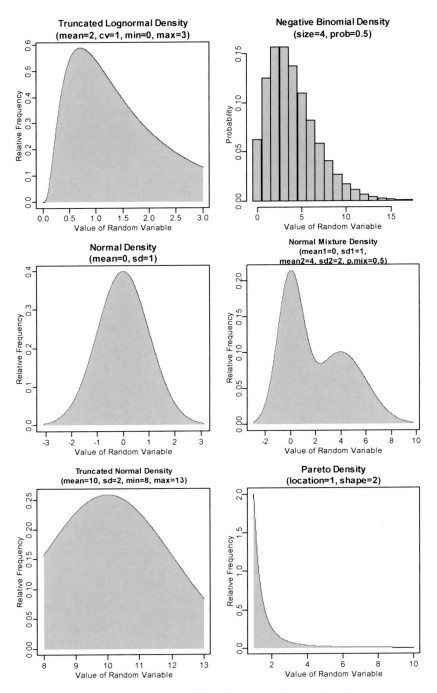

Fig. 4.1 (continued). Probability distributions in R and EnvStats

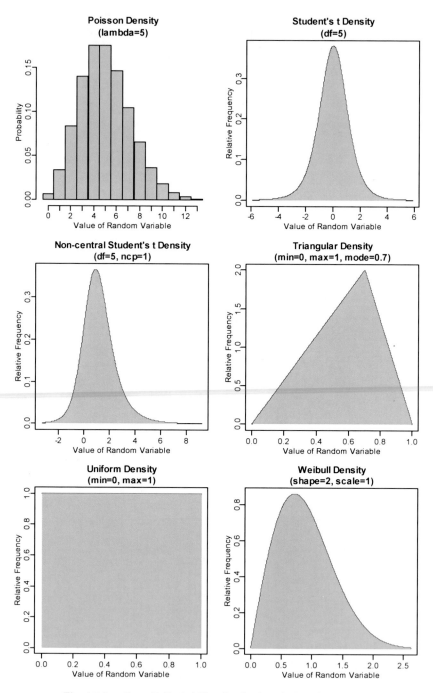

Fig. 4.1 (continued). Probability distributions in R and ENVSTATS

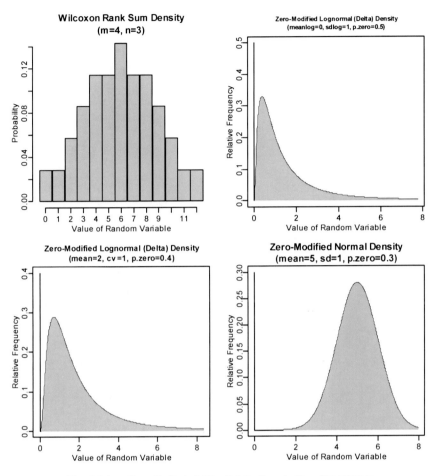

Fig. 4.1 (continued). Probability distributions in R and ENVSTATS

For each of these distributions, there are functions for computing the probability density function (pdf), the cumulative distribution function (cdf), quantiles, and random numbers. The form of the names of the functions are `dabb`, `pabb`, `qabb`, and `rabb`, where `abb` denotes the abbreviation of the distribution name (see column 2 of Table 4.1). For example, the functions `dnorm`, `pnorm`, `qnorm`, and `rnorm` compute the pdf, cdf, quantiles, and random numbers for the normal distribution.

Table 4.2 lists the functions available in R and ENVSTATS for plotting probability distributions, computing quantities associated with these distributions, and generating random numbers from these distributions.

Function(s)	Output
dabb, pabb, qabb	Probability density, cumulative distribution function, or quantiles for distribution with abbreviation *abb*
rabb simulateVector*	Random numbers from distribution with abbreviation *abb*
simulateMvMatrix*	Multivariate random numbers from various distributions based on user-specified rank correlation matrix
pdfPlot* epdfPlot* cdfPlot*	Plot of probability density function or cumulative distribution function

Table 4.2 Functions for probability distributions and random numbers (*part of ENVSTATS)

4.2   Probability Density Function (PDF)

A *probability density function (pdf)* is a mathematical formula that describes the relative frequency of a random variable. Sometimes the picture of this formula is called the pdf. If a random variable is discrete, its probability density function is sometimes called a *probability mass function*, since it shows the "mass" of probability at each possible value of the random variable.

4.2.1   Probability Density Function for Lognormal Distribution

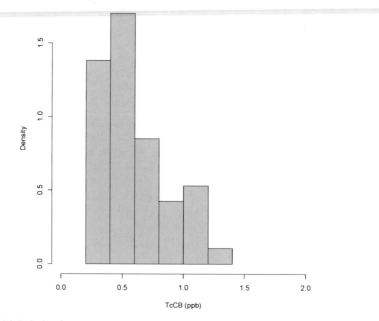

Fig. 4.2 Relative frequency (density) histogram of the Reference area TcCB data

Figure 4.2 shows the relative frequency (density) histogram for the Reference area TcCB data. It was created with these commands:

```
> with(EPA.94b.tccb.df ,
    hist(TcCB[Area == "Reference"], freq = FALSE,
    xlim = c(0, 2), xlab = "TcCB (ppb)",
    col = "cyan", main = ""))
```

For each class (bar) of this histogram, the proportion of observations falling in that class is equal to the area of the bar; that is, it is the width of the bar times the height of the bar. If we could take many, many more samples and create relative frequency histograms with narrower and narrower classes, we might end up with a picture that looks like Fig. 4.3 which shows the probability density function of a lognormal random variable with a mean of 0.6 and a coefficient of variation (CV) of 0.5. For a continuous random variable, a **probability distribution** can be thought of as what a density (relative frequency) histogram of outcomes would look like if you could keep taking more and more samples and making the histogram bars narrower and narrower. Figure 4.3 was created with this command:

```
> pdfPlot(distribution = "lnormAlt",
    param.list = list(mean = 0.6, cv = 0.5),
    curve.fill.col = "cyan", main = "")
```

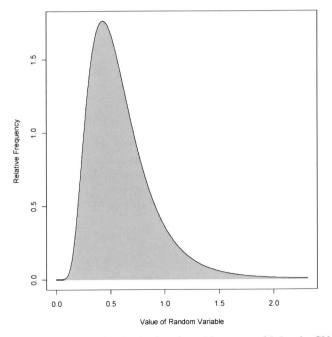

Fig. 4.3 Lognormal probability density function with a mean of 0.6 and a CV of 0.5

The probability density function for the lognormal distribution shown in Fig. 4.3 is given by:

$$f(x) = \frac{1}{x\sigma\sqrt{2\pi}}\, e^{\frac{-1}{2\sigma^2}\left[\log(x)-\mu\right]^2}, \quad x > 0 \tag{4.1}$$

where

$$\mu = \log\left(\frac{\theta}{\sqrt{\tau^2 + 1}}\right)$$

$$\sigma = \sqrt{\log\left(\tau^2 + 1\right)}$$

$$\theta = 0.6 \tag{4.2}$$

$$\tau = 0.5$$

and θ and τ denote the mean and coefficient of variation of the distribution, and μ and σ denote the mean and standard deviation of the log-transformed random variable. The values of the pdf evaluated at 0, 0.5, 1, 1.5, and 2, to three decimal places, are given by:

```
> round(dlnormAlt(seq(0, 2, by = 0.5), mean=0.6, cv=0.5),
    digits = 3)

[1] 0.000 1.670 0.355 0.053 0.009
```

4.2.2 Probability Density Function for a Gamma Distribution

As stated in Chap. 1, some EPA guidance documents (e.g., Singh et al. 2002; Singh et al. 2010a, b) discourage using the assumption of a lognormal distribution and recommend instead using a gamma distribution if it appears to fit the data. Figure 4.4 shows the probability density function of a gamma random variable with a mean of 0.6 and a coefficient of variation of 0.5 and was created with this command:

```
> pdfPlot(distribution = "gammaAlt",
    param.list = list(mean = 0.6, cv = 0.5), main = "")
```

The probability density function for the gamma distribution shown in Fig. 4.4 is given by:

$$f(x) = \frac{1}{\theta^k} \frac{1}{\Gamma(k)} x^{k-1} \ e^{-x/\theta} \ , \ x \geq 0; \ k, \theta > 0 \qquad (4.3)$$

where k and θ denote the shape and scale parameters, $\Gamma()$ denotes the gamma function, and

$$\mu = k\theta$$

$$\tau = 1/\sqrt{k}$$

$$\mu = 0.6 \qquad (4.4)$$

$$\tau = 0.5$$

where μ and τ denote the mean and coefficient of variation of the distribution. The values of the pdf evaluated at 0, 0.5, 1, 1.5, and 2, to three decimal places, are given by:

```
> round(dgammaAlt(seq(0, 2, by = 0.5), mean=0.6, cv=0.5),
    digits = 3)

[1] 0.000 1.468 0.419 0.050 0.004
```

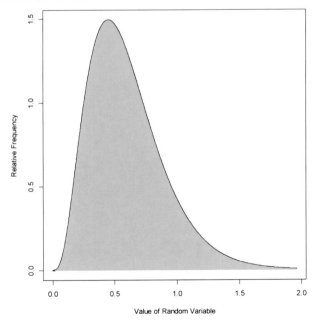

Fig. 4.4 Gamma probability density function with a mean of 0.6 and a CV of 0.5

4.3   Cumulative Distribution Function (CDF)

The *cumulative distribution function (cdf)* of a random variable X, sometimes called simply the distribution function, is the function F such that

$$F(x) = \Pr(X \le x) \tag{4.5}$$

for all values of x. That is, $F(x)$ is the probability that the random variable X is less than or equal to some number x. The cdf can also be defined or computed in terms of the probability density function (pdf) f as

$$F(x) = \Pr(X \le x) = \int_{-\infty}^{x} f(t)\, dt \tag{4.6}$$

4.3.1   Cumulative Distribution Function for Lognormal Distribution

Figure 4.5 displays the cumulative distribution function for the lognormal random variable whose pdf was shown in Fig. 4.3. It was created with this command:

```
> cdfPlot(distribution = "lnormAlt",
    param.list = list(mean = 0.6, cv = 0.5), main = "")
```

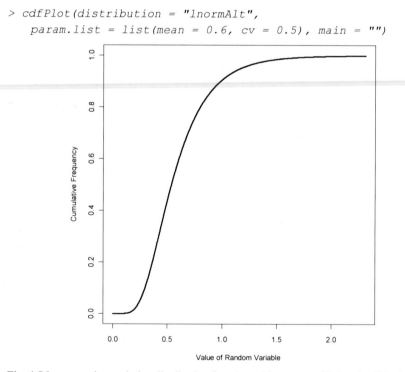

Fig. 4.5 Lognormal cumulative distribution function with a mean of 0.6 and a CV of 0.5

The values of the cdf evaluated at 0, 0.5, 1, 1.5, and 2, to two decimal places, are given by:

```
> round(plnormAlt(seq(0, 2, by = 0.5), mean=0.6, cv=0.5),
    digits = 2)
[1] 0.00 0.44 0.91 0.99 1.00
```

4.4 Quantiles and Percentiles

Loosely speaking, the *p*th **quantile** of a population is the (a) number such that a fraction *p* of the population is less than or equal to this number. The *p*th quantile is the same as the 100*p*th **percentile**; for example, the 0.5 quantile is the same as the 50th percentile.

Here is a more technical definition of a quantile. If X is a random variable with some specified distribution, the *p*th **quantile** of the distribution of X, denoted x_p, is a (the) number that satisfies:

$$\Pr\left(X < x_p\right) \le p \le \Pr\left(X \le x_p\right) \tag{4.7}$$

where *p* is a number between 0 and 1 (inclusive). If there is more than one number that satisfies the above condition, the *p*th quantile of X is often taken to be the average of the smallest and largest numbers that satisfy the condition. The R functions for computing quantiles, however, return the smallest number that satisfies the above condition.

If X is a continuous random variable, the *p*th quantile of X is simply defined as the value such that the cdf of that value is equal to *p*:

$$\Pr\left(X \le x_p\right) = F\left(x_p\right) = p \tag{4.8}$$

The 100*p*th **percentile** is another name for the *p*th quantile. That is, the 100*p*th percentile is the (a) number such that 100*p* % of the distribution lies below this number.

4.4.1 Quantiles for Lognormal Distribution

A plot of the cumulative distribution function makes it easy to visually pick out important quantiles, such as the median (50th percentile) or the 95th percentile. Looking at the cdf of the lognormal distribution shown in Fig. 4.5, the median (50th percentile) is about 0.5 and the 95th percentile is about 1.1. To compute the 50th and 95th percentiles of this lognormal distribution, type this command:

```
> qlnormAlt(c(0.5, 0.95), mean = 0.6, cv = 0.5)
[1] 0.5366563 1.1671907
```

4.5   Generating Random Numbers

With the advance of modern computers, experiments and simulations that just a couple of decades ago would have required an enormous amount of time to complete using large-scale computers can now be easily carried out on personal computers. Simulation is now an important tool in environmental statistics and all fields of statistics in general.

For all of the distributions shown in Fig. 4.1, you can generate random numbers (actually, pseudo-random numbers) from these distributions using R and ENVSTATS. Chapter 9 discusses how random numbers are generated in R and ENVSTATS, and how to use simulation to do environmental risk assessment.

4.5.1   Generating Random Numbers from a Univariate Distribution

To generate five random numbers from the lognormal distribution shown in Figs 4.3 and 4.5, type these commands:

```
> set.seed(23)

> rlnormAlt(5, mean = 0.6, cv = 0.5)

[1] 0.5879416 0.4370390 0.8261428 1.2520279 0.8593145
```

Note that you do not have to call the function set.seed before you generate random numbers. However, if you leave out the call to set.seed, the random numbers you generate will not be the same as the ones shown here.

4.5.2   Generating Multivariate Normal Random Numbers

In R you can generate random observations from a multivariate normal distribution using the function mvrnorm, which is part of the MASS package. Consider a bivariate normal distribution with the following parameters:

$$\mu = (\mu_1, \mu_2) = (5, 10)$$

$$\sigma = (\sigma_1, \sigma_2) = (1, 3) \tag{4.9}$$

$$\rho = \begin{pmatrix} 1 & 0.5 \\ 0.5 & 1 \end{pmatrix}$$

where μ denotes the vector of means, σ denotes the vector of standard deviations, and ρ denotes the correlation matrix. The function mvrnorm requires you to supply the covariance matrix, Σ, and the relationship between the covariance matrix, the standard deviations, and correlation matrix is given by:

$$\Sigma = \begin{pmatrix} \sigma_1 & 0 \\ 0 & \sigma_2 \end{pmatrix} \rho \begin{pmatrix} \sigma_1 & 0 \\ 0 & \sigma_2 \end{pmatrix} \qquad (4.10)$$

To generate three random observations from this bivariate distribution, type these commands:

```
> library(MASS)

> set.seed(47)

> sd.vec <- c(1, 3)

> cor.mat <- matrix(c(1, 0.5, 0.5, 1), ncol = 2)

> cor.mat

     [,1] [,2]
[1,]  1.0  0.5
[2,]  0.5  1.0

> cov.mat <- diag(sd.vec) %*% cor.mat %*% diag(sd.vec)

> cov.mat

     [,1] [,2]
[1,]  1.0  1.5
[2,]  1.5  9.0

> mvrnorm(n = 3, mu = c(5, 10), Sigma = cov.mat)

         [,1]     [,2]
[1,] 5.847172 16.01927
[2,] 5.477687 12.11412
[3,] 4.189215 10.72079
```

4.5.3 Generating Multivariate Observations Based on Rank Correlations

In ENVSTATS, you can use the function `simulateMvMatrix` to generate multivariate correlated observations where each variable has an arbitrary distribution. For example, you can generate a multivariate observation ($X1$, $X2$) where $X1$ comes from a normal distribution and $X2$ comes from a lognormal distribution. As an example, suppose $X1$ follows a normal distribution with mean 5 and standard deviation 1, $X2$ follows a lognormal distribution with mean 10 and CV 2, and we desire a rank correlation specified by ρ in Eq. 4.9. Here is a command to generate three random observations from this bivariate distribution:

```
> simulateMvMatrix(n = 3,
    distributions = c(X1 = "norm", X2 = "lnormAlt"),
    param.list = list(X1 = list(mean = 5, sd = 1),
                      X2 = list(mean = 10, cv = 2)),
    cor.mat = matrix(c(1, 0.5, 0.5, 1), ncol=2), seed = 105)

          X1          X2
[1,]  5.117840  1.595663
[2,]  5.404491  7.100144
[3,]  6.294479  6.055441
```

4.6 Summary

- Figure 4.1 displays examples of the probability density functions for the probability distributions available in R and ENVSTATS. Many of these distributions are already available in R, and some have been added in ENVSTATS.

- Table 4.1 lists the probability distributions available in R and ENVSTATS, along with their abbreviations and associated parameters. For each of these distributions, you can compute the probability density function (pdf), the cumulative distribution function (cdf), quantiles, and random numbers. You can also plot the pdf and/or cdf.

- Table 4.2 lists the functions available in ENVSTATS for plotting probability distributions, computing quantities associated with these distributions, and generating random numbers from these distributions.

Chapter 5

Estimating Distribution Parameters and Quantiles

5.1 Introduction

In Chap. 2 we discussed the ideas of a population and a sample. Chapter 4 described probability distributions, which are used to model populations. Based on using the graphical tools discussed in Chap. 3 to look at your data, and based on your knowledge of the mechanism producing the data, you can model the data from your sampling program as having come from a particular kind of probability distribution. Once you decide on what probability distribution to use (if any), you usually need to estimate the parameters associated with that distribution. For example, you may need to compare the mean or 95th percentile of the concentration of a chemical in soil, groundwater, surface water, or air with some fixed standard. This chapter discusses the functions available in ENVSTATS for estimating distribution parameters and quantiles for various probability distributions, as well as constructing confidence intervals (CIs) for these quantities. See Millard et al. (2014) for a more in-depth discussion of this topic.

5.2 Estimating Distribution Parameters

Table 4.1 lists the probability distributions available in R and ENVSTATS. For most of these distributions, there are EnvStats functions for estimating the parameters of these distributions (see the EnvStats help file *Estimating Distribution Parameters* for a complete list). The form of the names of these functions is e*abb*, where *abb* denotes the abbreviation of the distribution name (see column 2 of Table 4.1). For example, the function enorm estimates the mean and standard deviation based on a set of observations assumed to come from a normal distribution, and also optionally allows you to construct a confidence interval for the mean or variance.

5.2.1 Estimating Parameters of a Normal Distribution

Recall that in Chap. 1 we saw that the Reference area TcCB data appeared to come from a lognormal distribution. Here is the estimated mean and standard deviation of the log-transformed data, along with a 95 % confidence interval for the mean:

```
> attach(EPA.94b.tccb.df)

> enorm.list <- enorm(log(TcCB[Area == "Reference"]),
    ci = TRUE))

> enorm.list
```

Results of Distribution Parameter Estimation

Assumed Distribution:	Normal
Estimated Parameter(s):	mean = -0.6195712
	sd = 0.4679530
Estimation Method:	mvue
Data:	log(TcCB[Area == "Reference"])
Sample Size:	47
Confidence Interval for:	mean
Confidence Interval Method:	Exact
Confidence Interval Type:	two-sided
Confidence Level:	95%
Confidence Interval:	LCL = -0.7569673
	UCL = -0.4821751

Note that calling the function enorm with the log-transformed data gives the
same results as calling the function elnorm with the untransformed data
(see Sect. 1.11.6). Figure 5.1 shows a density histogram of the log-transformed
Reference area TcCB data, along with the fitted normal distribution based on these
estimates. It was create with these commands:

```
> hist(log(TcCB[Area == "Reference"]), freq = FALSE,
    xlim = c(-2, 1), xlab = "log [ TcCB (ppb) ]",
    ylim = c(0, 1), col = "cyan", main = "")

> params <- enorm.list$parameters

> pdfPlot(dist = "norm", param.list = as.list(params),
    add = TRUE)
```

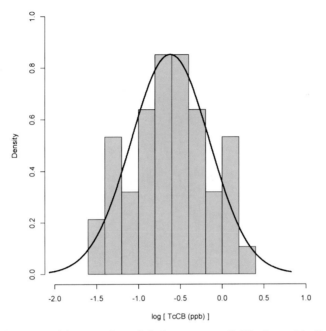

Fig. 5.1 Histogram of log-transformed Reference area TcCB data with fitted normal distribution

5.2.2 *Estimating Parameters of a Lognormal Distribution*

Rather than estimate parameters based on the log-transformed TcCB Reference area data, we can estimate the parameters of the lognormal distribution based on the original scale. For the untransformed Reference area TcCB data, the estimated mean is 0.6 ppb and the estimated coefficient of variation is 0.49. Figure 5.2 shows a density histogram of the Reference area TcCB data, along with the fitted lognormal distribution based on these estimates. The two-sided 95 % confidence interval for the mean based on Land's method is [0.52, 0.70] ppb.

```
> elnormAlt.list <- elnormAlt(TcCB[Area == "Reference"],
    ci = TRUE)

> elnormAlt.list

Results of Distribution Parameter Estimation
--------------------------------------------

Assumed Distribution:          Lognormal

Estimated Parameter(s):        mean = 0.5989072
                               cv   = 0.4899539
```

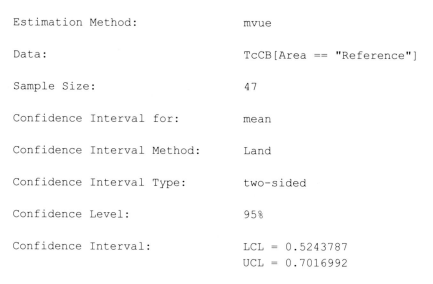

```
Estimation Method:                mvue

Data:                             TcCB[Area == "Reference"]

Sample Size:                      47

Confidence Interval for:          mean

Confidence Interval Method:       Land

Confidence Interval Type:         two-sided

Confidence Level:                 95%

Confidence Interval:              LCL = 0.5243787
                                  UCL = 0.7016992
```

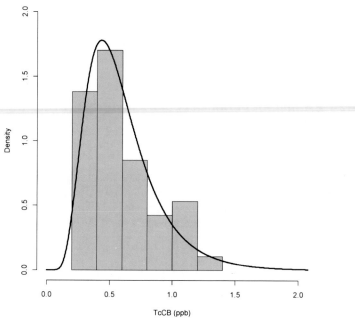

Fig. 5.2 Histogram of Reference area TcCB data with fitted lognormal distribution

To create Fig. 5.2, type these commands:

```
> hist(TcCB[Area == "Reference"], freq = FALSE,
    xlim = c(0, 2), xlab = "TcCB (ppb)", ylim = c(0, 2),
    col = "cyan", main = "")
```

```
> params <- elnormAlt.list$parameters

> pdfPlot(dist = "lnormAlt", param.list = as.list(params),
    add = TRUE)
```

By default, the function `elnormAlt` uses the method of Land (1971, 1975) to compute a confidence interval for the mean of a lognormal distribution. Although Land's method is exact (i.e., the confidence level is exact and does not depend on asymptotic theory), it is computationally intensive. Zou et al. (2009) present a simpler alternative method that appears to perform quite well even for small sample sizes. They give an example using data on ambient air lead levels ($\mu g/m^3$) collected by the National Institute of Occupation Safety and Health (NIOSH) in 1989 that appear in Krishnamoorthy et al. (2006).

```
> sort(NIOSH.89.air.lead.vec)
```

```
[1]    6    7    8   15   29   48   61   80  110  120  200
[12]  350  380 1000 1400
```

```
> round(elnormAlt(NIOSH.89.air.lead.vec,
    ci = TRUE)$interval$limits)
```

```
LCL   UCL
 117  4038
```

```
> round(elnormAlt(NIOSH.89.air.lead.vec, ci = TRUE,
    ci.method = "zou")$interval$limits)
```

```
LCL   UCL
 112  3873
```

Neither of these methods may prove satisfactory for small sample sizes because the upper confidence limit can be much larger (e.g., an order of magnitude larger) than the largest observation. Some authors (e.g., USEPA 1997d) erroneously claim that using the bootstrap can overcome this problem. In fact, confidence intervals for the mean of a lognormal distribution based on the bootstrap fail to provide adequate coverage for small sample sizes (Millard et al. 2014).

5.2.3   Estimating Parameters of a Gamma Distribution

One way to avoid the problem of potentially large confidence limits associated with an assumed lognormal distribution is to not assume a lognormal distribution at all, but instead assume a gamma distribution if it makes sense, as recommended by the EPA guidance documents Singh et al. (2002) and Singh et al. (2010a, b). In Chap. 1 we saw that the gamma distribution appeared to be an adequate model for the Reference area TcCB data (see Figs. 1.9 and 1.11). Using the `egamma` and `egammaAlt` functions (Sect. 1.11.6), the estimated shape and scale parameters are 4.9 and 0.1, respectively, the estimated mean and CV are 0.6 and 0.45, respectively, and the 95 % confidence interval for the mean is [0.52, 0.68] ppb, compared with [0.52, 0.70] based on the assumption of a lognormal distribution. By default, the confidence interval for the mean is based on the method of Kulkarni and Powar (2010), which involves approximating the gamma distribution with a normal distribution. See Millard et al. (2014) or the help files for details.

5.2.4   Estimating the Parameter of a Binomial Distribution

The guidance document *Statistical Analysis of Ground-Water Monitoring Data at RCRA Facilities: Addendum to Interim Final Guidance* (USEPA 1992c) contains observations on benzene concentrations (ppb) in groundwater from six background wells sampled monthly for 6 months. The data are stored in the data frame `EPA.92c.benzene1.df` in ENVSTATS. Nondetect values are reported as "<2" and of the 36 values, 33 are nondetects. Section 2.8.2 in Chap. 2 showed how to use the EnvStats function `ebinom` to estimate the probability of observing a nondetect value at any of the six wells as about 92 %, with the two-sided 95 % confidence interval for the binomial proportion based on using the normal score approximation with continuity correction as [76 %, 98 %].

5.3   Estimating Distribution Quantiles

We defined the *p*th quantile or the 100*p*th percentile of a distribution in Sect. 4.4 of Chap. 4. Quantiles or percentiles are sometimes used in environmental standards and regulations (e.g., Berthouex and Brown 2002). For example, in order to determine compliance, you may be required to estimate an extreme percentile (e.g., the 95th percentile) for the "background level" distribution, and then compare observations at compliance wells or remediated areas to this upper percentile (or an upper confidence limit for this percentile). In the context of soil cleanup, USEPA (1994b) has called this the "Hot-Measurement Comparison." (There are some major problems with this technique that are discussed in Millard et al. 2014.)

As another example, when monitoring groundwater around a RCRA landfill, the site may be in compliance/assessment or corrective action monitoring for a particular chemical constituent, where data are compared to a groundwater protection standard (GWPS). "In compliance/assessment, the comparison is made to determine whether groundwater concentrations have increased above the compliance standard. In corrective action, the test determines whether concentrations have decreased below a clean-up criterion or compliance level. In compliance/assessment monitoring, the lower confidence limit [LCL] is of primary interest, while the upper confidence limit [UCL] is most important in corrective action," (USEPA 2009, pp. 21–1). The fixed compliance limit may be a maximum concentration limit (MCL) or an alternate concentration limit (ACL). Most MCLs and ACLs appear to represent long-term average levels, but sometimes they may represent a limit that should be exceeded only a small fraction of the time, for example, the 95th percentile of the distribution. In this case you need to compare the 95th percentile of the distribution of the chemical's concentration in the groundwater with the GWPS. Under compliance/assessment monitoring, if the *lower* confidence limit for the specified percentile is *greater* than the GWPS, this indicates the facility is out of compliance for that constituent. Under corrective action monitoring, if the *upper* confidence limit for the specified percentile is *less* than the GWPS, this indicates the facility appears to have cleaned up the contamination and should be able to return to compliance/assessment monitoring.

In EnvStats, functions for estimating quantiles have names of the form eqabb, where *abb* denotes the abbreviation of the distribution name (see column 2 of Table 4.1). Some of these functions let you create confidence intervals for quantiles as well. You can also estimate quantiles and create confidence intervals for them nonparametrically using the function eqnpar. See the EnvStats help file *Estimating Distribution Quantiles* for a complete list of functions.

5.3.1   *Estimating Quantiles of a Normal Distribution*

The guidance document *Statistical Analysis of Groundwater Monitoring Data at RCRA Facilities: Unified Guidance* (USEPA 2009) contains an example on page 21–13 where aldicarb concentrations (ppb) at three compliance wells (four monthly samples at each well) are to be compared against an MCL of 30 ppb. The MCL should not be exceeded more than 5 % of the time. The data for this example are stored in EPA.09.Ex.21.1.aldicarb.df.

```
> EPA.09.Ex.21.1.aldicarb.df
```

	Month	Well	Aldicarb.ppb
1	1	Well.1	19.9
2	2	Well.1	29.6
3	3	Well.1	18.7
4	4	Well.1	24.2
5	1	Well.2	23.7
6	2	Well.2	21.9
7	3	Well.2	26.9
8	4	Well.2	26.1
9	1	Well.3	5.6
10	2	Well.3	3.3
11	3	Well.3	2.3
12	4	Well.3	6.9

First we assume the facility is in compliance/assessment monitoring, so we are instructed to compute a one-sided *lower* 99 % confidence limit for the 95th percentile for each well and compare that to the GWPS. Here are the results for Well 1:

```
> attach(EPA.09.Ex.21.1.aldicarb.df)

> eqnorm(Aldicarb.ppb[Well == "Well.1"], p = 0.95, ci = TRUE,
    ci.type = "lower", conf.level = 0.99)
```

```
Results of Distribution Parameter Estimation
--------------------------------------------

Assumed Distribution:           Normal

Estimated Parameter(s):         mean = 23.10000
                                sd   =  4.93491

Estimation Method:              mvue

Estimated Quantile(s):          95'th %ile = 31.21720

Quantile Estimation Method:     qmle

Data:                           Aldicarb.ppb[Well == "Well.1"]

Sample Size:                    4

Confidence Interval for:        95'th %ile

Confidence Interval Method:     Exact

Confidence Interval Type:       lower

Confidence Level:               99%

Confidence Interval:            LCL = 25.28550
                                UCL =      Inf
```

To compute the LCL for all three wells at once, type this command:

```
> sapply(split(Aldicarb.ppb, Well), function(x) {
    eqnorm(x, p = 0.95, ci = TRUE, ci.type = "lower",
    conf.level = 0.99)$interval$limits["LCL"]})

Well.1.LCL Well.2.LCL Well.3.LCL
  25.28550   25.66086    5.45563
```

Since none of the LCLs is above 30 ppb, no corrective action is needed. On the other hand, if we assume the site is in corrective action monitoring, then we need to compute the one-sided 99 % *upper* confidence limit for the 95th percentile and compare that to the MCL:

```
> sapply(split(Aldicarb.ppb, Well), function(x) {
    eqnorm(x, p = 0.95, ci = TRUE, ci.type = "upper",
    conf.level = 0.99)$interval$limits["UCL"]})
```

```
Well.1.UCL Well.2.UCL Well.3.UCL
  67.92601    45.38336    23.61286
```

In this case there is evidence that corrective action is still needed at Wells 1 and 2 since the UCL is greater than 30 ppb. Of course, a major consideration in this whole example is the very small sample size at each well ($n = 4$) used to compute the intra-well confidence limit.

5.3.2   Estimating Quantiles of a Lognormal Distribution

The guidance document *Statistical Analysis of Groundwater Monitoring Data at RCRA Facilities: Unified Guidance* (USEPA 2009) contains an example on page 17–17 of chrysene concentrations (ppb) from groundwater monitoring at two background wells and three compliance wells (four monthly samples at each well). In ENVSTATS these data are stored in EPA.09.Ex.17.3.chrysene.df.

```
> EPA.09.Ex.17.3.chrysene.df
```

```
   Month   Well  Well.type Chrysene.ppb
1      1 Well.1 Background        19.7
2      2 Well.1 Background        39.2
3      3 Well.1 Background         7.8
4      4 Well.1 Background        12.8
5      1 Well.2 Background        10.2
6      2 Well.2 Background         7.2
7      3 Well.2 Background        16.1
8      4 Well.2 Background         5.7
9      1 Well.3 Compliance       68.0
10     2 Well.3 Compliance       48.9
11     3 Well.3 Compliance       30.1
12     4 Well.3 Compliance       38.1
13     1 Well.4 Compliance       26.8
14     2 Well.4 Compliance       17.7
15     3 Well.4 Compliance       31.9
16     4 Well.4 Compliance       22.2
17     1 Well.5 Compliance       47.0
18     2 Well.5 Compliance       30.5
19     3 Well.5 Compliance       15.0
20     4 Well.5 Compliance       23.4
```

In this example, we compute a 95 % upper confidence limit for the 95th percentile based on the data from the two background wells, and compare all of the observations at the three compliance wells to this UCL to determine whether any of them are out of compliance. The example in the guidance document shows that a lognormal distribution appears to fit these data.

```
> attach(EPA.09.Ex.17.3.chrysene.df)
```

```
> Chrysene <- Chrysene.ppb[Well.type == "Background"]
> eqlnorm(Chrysene, p = 0.95, ci = TRUE, ci.type = "upper",
    conf.level = 0.95)
```

```
Results of Distribution Parameter Estimation
--------------------------------------------

Assumed Distribution:                   Lognormal

Estimated Parameter(s):                 meanlog = 2.5085773
                                        sdlog   = 0.6279479

Estimation Method:                      mvue

Estimated Quantile(s):                  95'th %ile = 34.51727

Quantile Estimation Method:             qmle

Data:                                   Chrysene

Sample Size:                            8

Confidence Interval for:                95'th %ile

Confidence Interval Method:             Exact

Confidence Interval Type:               upper

Confidence Level:                       95%

Confidence Interval:                    LCL =  0.0000
                                        UCL = 90.9247
```

Since none of the observations from the compliance wells exceed the UCL of 90.9 ppb, there is no evidence of contamination. Note that although this example in the EPA guidance document is presented in the section on tolerance intervals in that document, tolerance intervals (specifically β-content tolerance intervals) and confidence intervals for percentiles are the same thing; see Millard et al. (2014) for details.

5.3.3   Estimating Quantiles of a Gamma Distribution

Instead of assuming the chrysene data in the previous section comes from a lognormal distribution, we could instead assume it comes from a gamma distribution.

In this case, the estimated 95 % UCL of the 95th percentile is only 69.3 ppb instead of 90.9, almost a 25 % reduction in the UCL!

```
> eqgamma(Chrysene, p = 0.95, ci = TRUE, ci.type = "upper",
    conf.level = 0.95)
```

```
Results of Distribution Parameter Estimation
--------------------------------------------


Assumed Distribution:            Gamma

Estimated Parameter(s):          shape = 2.806929
                                 scale = 5.286026

Estimation Method:               mle

Estimated Quantile(s):           95'th %ile = 31.74348

Quantile Estimation Method:      Quantile(s) Based on
                                 mle Estimators

Data:                            Chrysene

Sample Size:                     8

Confidence Interval for:         95'th %ile

Confidence Interval Method:      Exact using
                                 Kulkarni & Powar (2010)
                                 transformation to Normality
                                 based on mle of 'shape'

Confidence Interval Type:        upper

Confidence Level:                95%

Confidence Interval:             LCL =   0.00000
                                 UCL = 69.32425
```

5.3.4 *Nonparametric Estimates of Quantiles*

To estimate quantiles nonparametrically, all you need to do is estimate the cdf nonparametrically using the empirical cdf, then use linear interpolation (if necessary). Graphically, this just means connecting the points in the quantile plot by straight lines, finding the value of p on the y-axis, and determining the corresponding number on the x-axis. For example, looking at Fig. 1.4 in Sect. 1.11.4, we might estimate the median (i.e., $p = 0.5$) of the Reference area TcCB data to be about 0.5 ppb (the actual value is 0.54 ppb).

One problem with estimating quantiles nonparametrically versus parametrically is that you need many more observations to estimate extreme quantiles with good precision. In fact, even though the `quantile` function in R allows you to estimate any quantile for any sample size, in the case where say $n = 10$, it does not make sense intuitively that we should be able to estimate anything less than the 10th percentile or anything more than the 90th percentile with any kind of precision. This characteristic becomes clear when we create nonparametric confidence intervals for quantiles. On the other hand, an advantage to estimating quantiles nonparametrically is that it is often easy to deal with censored values since all you have to do is rank them.

Nonparametric confidence intervals for quantiles are based on the ranked data, and usually the largest value is used for the upper confidence limit for a large percentile, and the smallest value is used for the lower confidence limit for a small percentile. The confidence level associated with these confidence intervals depends on the sample size. For example, to compute the confidence levels associated with a one-sided upper confidence interval for the 95th percentile based on various sample sizes assuming the upper confidence limit is the maximum value, type these commands:

```
> Sample.Size <- c(seq(5, 25, by = 5), 50, 75, 100)

> conf.level <- tolIntNparConfLevel(Sample.Size,
    coverage = 0.95, ti.type = "upper")

> cbind(Sample.Size,
    Confidence.Level = round(100 * conf.level))
```

	Sample.Size	Confidence.Level
[1,]	5	23
[2,]	10	40
[3,]	15	54
[4,]	20	64
[5,]	25	72
[6,]	50	92
[7,]	75	98
[8,]	100	99

You can see that a confidence level greater than 95 % cannot be achieved until the sample size is larger than $n = 50$. See Millard et al. (2014) for a detailed discussion of estimating quantiles nonparametrically and constructing nonparametric confidence intervals for quantiles.

The guidance document *Statistical Analysis of Groundwater Monitoring Data at RCRA Facilities: Unified Guidance* (USEPA 2009) contains an example on page 21–21 where nitrate concentrations (mg/L) at a well that is used for drinking water are to be compared against the infant-based, acute risk standard of 10 mg/L.

The risk standard represents the upper 95th percentile limit on nitrate concentrations and we want to be 95 % confident that the risk standard has not been violated. First we assume the facility is in compliance/assessment monitoring, so we are instructed to compute one-sided *lower* 95 % confidence limit for the 95th percentile and compare it to the risk standard of 10 mg/L. The data for this example are stored in EPA.09.Ex.21.6.nitrate.df.

```
> EPA.09.Ex.21.6.nitrate.df[, 1:3]

      Sampling.Date        Date Nitrate.mg.per.l.orig
1        7/28/1999 1999-07-28                    <5.0
2         9/3/1999 1999-09-03                    12.3
3       11/24/1999 1999-11-24                    <5.0
4         5/3/2000 2000-05-03                    <5.0
5        7/14/2000 2000-07-14                     8.1
6       10/31/2000 2000-10-31                    <5.0
7       12/14/2000 2000-12-14                      11
8        3/27/2001 2001-03-27                    35.1
9        6/13/2001 2001-06-13                    <5.0
10       9/16/2001 2001-09-16                    <5.0
11      11/26/2001 2001-11-26                     9.3
12       3/2/2002 2002-03-02                    10.3
```

Because the data contain censored observations, two additional columns were added to indicate the numeric value and whether or not the observation was censored:

```
> EPA.09.Ex.21.6.nitrate.df[, c(2, 4:5)]

         Date Nitrate.mg.per.l Censored
1  1999-07-28              5.0     TRUE
2  1999-09-03             12.3    FALSE
3  1999-11-24              5.0     TRUE
4  2000-05-03              5.0     TRUE
5  2000-07-14              8.1    FALSE
6  2000-10-31              5.0     TRUE
7  2000-12-14             11.0    FALSE
8  2001-03-27             35.1    FALSE
9  2001-06-13              5.0     TRUE
10 2001-09-16              5.0     TRUE
11 2001-11-26              9.3    FALSE
12 2002-03-02             10.3    FALSE
```

For this data set, half of the values are nondetects, so estimating the median nonparametrically is problematic and estimating percentiles less than 50 % nonparametrically is not possible. We need to determine which ranked value to use for the lower confidence limit for the 95th percentile in order to achieve at least 95 % confidence.

```
> Nitrate <- EPA.09.Ex.21.6.nitrate.df$ Nitrate.mg.per.l

> eqnpar(Nitrate, p = 0.95, ci = TRUE, ci.type = "lower",
    approx.conf.level = 0.95)

Results of Distribution Parameter Estimation
--------------------------------------------

Assumed Distribution:              None

Estimated Quantile(s):             95'th %ile = 22.56

Quantile Estimation Method:        Nonparametric

Data:                              Nitrate

Sample Size:                       12

Confidence Interval for:           95'th %ile

Confidence Interval Method:        exact

Confidence Interval Type:          lower

Confidence Level:                  88%

Confidence Limit Rank(s):          11

Confidence Interval:               LCL = 12.3
                                   UCL =  Inf
```

In this example, by default, the **EnvStats** function eqnpar uses the 11th largest value (12.3 mg/L) as the lower confidence limit, but this yields only an 88 % confidence level. Using the 10th largest value yields a confidence level of 98 %:

```
> eqnpar(Nitrate, p = 0.95, ci = TRUE, ci.type = "lower",
    lcl.rank = 10)

Results of Distribution Parameter Estimation
--------------------------------------------

Assumed Distribution:              None

Estimated Quantile(s):             95'th %ile = 22.56
```

```
Quantile Estimation Method:        Nonparametric

Data:                              Nitrate

Sample Size:                       12

Confidence Interval for:           95'th %ile

Confidence Interval Method:        exact

Confidence Interval Type:          lower

Confidence Level:                  98%

Confidence Limit Rank(s):          10

Confidence Interval:               LCL =   11
                                   UCL = Inf
```

Because the 10th largest value is 11 mg/L and this is larger than the acute risk standard of 10 mg/L, we conclude there is evidence of contamination at the well.

If we assume the well was being remediated under corrective action monitoring, the fixed standard would be compared against a one-sided *upper* confidence limit for the 95th percentile. With a sample size of $n = 12$, using the largest value as the upper confidence limit yields a confidence level of only 46 %:

```
> eqnpar(Nitrate, p = 0.95, ci = TRUE, ci.type = "upper",
    approx.conf.level = 0.95)

Results of Distribution Parameter Estimation
--------------------------------------------

Assumed Distribution:              None

Estimated Quantile(s):             95'th %ile = 22.56

Quantile Estimation Method:        Nonparametric

Data:                              Nitrate

Sample Size:                       12

Confidence Interval for:           95'th %ile

Confidence Interval Method:        exact
```

```
Confidence Interval Type:              upper

Confidence Level:                      46%

Confidence Limit Rank(s):              12

Confidence Interval:                   LCL = -Inf
                                       UCL = 35.1
```

In order to achieve a confidence level of 95 %, we would need to have $n = 59$ observations and all of the observations would need to be less than the fixed standard of 10 mg/L in order for the well to return to compliance/assessment monitoring:

```
> tolIntNparN(coverage = 0.95, ti.type = "upper",
    conf.level = 0.95)

[1] 59
```

5.4 Summary

- Whether you are conducting a preliminary, descriptive study of the environment or monitoring the environment for contamination under a specific regulation, you usually need to characterize the distribution of whatever you are looking at (e.g., a chemical in the environment), which involves **estimating distribution parameters** such as the mean, median, standard deviation, 95th percentile, etc.
- You can use EnvStats functions of the form eabb (where abb denotes the abbreviation of the distribution name) for estimating distribution parameters and optionally constructing confidence intervals. These functions are listed in the help file *Estimating Distribution Parameters.*
- Functions for estimating quantiles and optionally constructing confidence intervals for them have names of the form eqabb. These functions are listed in the help file *Estimating Distribution Quantiles.*
- One problem with estimating quantiles nonparametrically versus parametrically is that you need many more observations to estimate extreme quantiles with good precision.

Chapter 6

Prediction and Tolerance Intervals

6.1 Introduction

Any activity that requires constant monitoring over time and the comparison of new values to "background" or "standard" values creates a decision problem: if the new values greatly exceed the background values, has a change really occurred, or have the true underlying concentrations stayed the same and this is just a "chance" event? Statistical tests are used as objective tools to decide whether a change has occurred (although the choice of Type I error level and acceptable power are subjective decisions). For a monitoring program that involves numerous tests over time, figuring out how to balance the *overall* Type I error with the power of detecting a change is not a trivial problem, but it is also a problem that has been dealt with for a long time in the statistical literature under the heading of "multiple comparisons." Prediction intervals and tolerance intervals are two tools that you can use to attempt to solve the multiple comparisons problem. This chapter discusses the functions available in ENVSTATS for constructing prediction and tolerance intervals. See Millard et al. (2014) for a more in-depth discussion of this topic.

6.2 Prediction Intervals

A *prediction interval* for some population is an interval on the real line constructed so that it will contain k future observations or averages from that population with some specified probability $(1-\alpha)100\,\%$, where α is some fraction between 0 and 1 (usually α is less than 0.5), and k is some pre-specified positive integer. Just as for confidence intervals, the quantity $(1-\alpha)100\,\%$ is called the *confidence coefficient* or *confidence level* associated with the prediction interval. Table 6.1 lists the functions available in ENVSTATS for constructing prediction intervals.

The basic idea of a prediction interval is to assume a particular probability distribution (e.g., normal, lognormal, etc.) for some process generating the data (e.g., quarterly observations of chemical concentrations in groundwater), compute sample statistics from a baseline sample, and then use these sample statistics to construct a prediction interval, assuming the distribution of the data does not change in the future (or if we are comparing one geographical area to another, we assume the distribution of data from the comparison area is the same as the distribution of data from the baseline area). If the future observation or observations do not fall within the prediction interval, then this is evidence that the distribution has potentially changed (e.g., contamination is present). For example, if X denotes a random variable from some population, and we know what the population looks like,

S.P. Millard, *EnvStats: An R Package for Environmental Statistics*,
DOI 10.1007/978-1-4614-8456-1_6, © Springer Science+Business Media New York 2013

Distribution	Function name	Description
Normal	predIntNorm	Construct a **prediction interval** for the next k observations or next k means from a normal distribution
	predIntNormK	Compute the value of K for a prediction interval for a normal distribution
	predIntNormSimultaneous	Construct a **simultaneous prediction interval** for the next r sampling occasions based on a normal distribution
	predIntNormSimultaneousK	Compute the value of K for a simultaneous prediction interval for the next r sampling occasions based on a normal distribution
Lognormal	predIntLnorm predIntLnormAlt	Construct a **prediction interval** based on a lognormal distribution
	predIntLnormSimultaneous predIntLnormAltSimultaneous	Construct a **simultaneous prediction interval** based on a lognormal distribution
Gamma	predIntGamma predIntGammaAlt	Construct a **prediction interval** based on a gamma distribution
	predIntGammaSimultaneous predIntGammaSimultaneousAlt	Construct a **simultaneous prediction interval** based on a gamma distribution
Poisson	predIntPois	Construct a prediction interval for the next k observations or sums from a Poisson distribution
Nonparametric	predIntNpar	Construct a nonparametric prediction interval for the next k of m observations

Table 6.1 Functions in EnvStats for constructing prediction intervals

(e.g., lognormal with a mean of 10 and a CV of 1), so we can compute the quantiles of the population, then a $(1-\alpha)100\,\%$ two-sided prediction interval for the next $k = 1$ observation of X is given by:

$$\left[x_{\alpha/2} , x_{1-\alpha/2} \right]$$
(6.1)

where x_p denotes the pth quantile of the distribution of X. Similarly, a $(1-\alpha)100\,\%$ one-sided *upper* prediction interval for the next observation is given by:

$$\left[-\infty , x_{1-\alpha} \right]$$
(6.2)

and a $(1-\alpha)100\,\%$ one-sided *lower* prediction interval for the next observation is given by:

$$\left[x_\alpha , \infty \right]$$
(6.3)

See Millard et al. (2014) for the corresponding equations for general values of k.

Usually the true distribution of X is unknown, so the values of the prediction limits have to be estimated based on estimating the parameters of the distribution of X. For the usual case when the exact distribution of X is unknown, a prediction interval is thus a random interval; that is, the lower and upper bounds are random variables computed based on sample statistics in the baseline sample. Prior to taking one specific baseline sample, the probability that the prediction interval *will* contain the next k observations is $(1-\alpha)100\%$. Once a specific baseline sample is taken and the prediction interval based on that sample is computed, the probability that that prediction interval will contain the next k observations is not necessarily $(1-\alpha)100\%$, but it should be close to this value for a moderately large sample size.

Suppose an experiment is performed N times, and suppose that for each experiment:

1. A sample is taken and a $(1-\alpha)100\%$ prediction interval for $k = 1$ future observation is computed.
2. One future observation is generated and compared to the prediction interval.

Then the number of times a prediction interval generated in Step 1 above will contain a future observation generated in step 2 above is a binomial random variable with parameters $n = N$ and $p = 1-\alpha$, that is, it follows a B(N, $1-\alpha$) distribution.

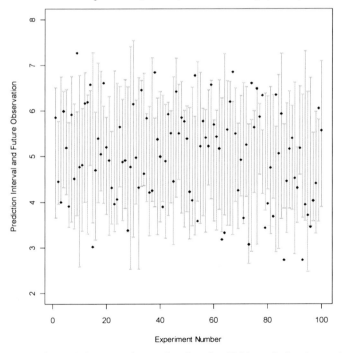

Fig. 6.1 Results of simulation experiment showing the 80 % prediction interval and one future observation for 100 simulations

Figure 6.1 shows the results of such a simulated experiment in which a random sample of $n = 10$ observations was taken from a N(5, 1) distribution and an 80 % prediction interval for $k = 1$ future observation was constructed based on these 10 observations. Then one future observation was generated. The experiment was repeated 100 times. In this case, the actual number of times the prediction interval contained the future observation was 79.

It is important to note that if only one baseline sample is taken and only one prediction interval for $k = 1$ future observation is computed, then the number of future observations out of a total of N future observations that will be contained in that one prediction interval is a binomial random variable with parameters $n = N$ and $p = 1-\alpha^*$, where α^* depends on the true population parameters and the computed bounds of the single prediction interval. For example, if we compute a prediction interval for $k = 1$ future observation, assuming the data used to create the prediction interval come from a N(5, 1) distribution and the prediction interval is [2.54, 7.25], then the total number of N future observations that will be contained in this prediction interval is a Binomial random variable with parameters $n = N$ and $p = 0.98$ since the probability that a single observation from a N(5, 1) distribution will fall in the interval [2.54, 7.25] is 98 %.

A prediction interval is usually constructed as a bound on future individual observations, but it can also be formulated as a bound on the *mean* of p future observations (or a bound on multiple future means). In a testing scenario, the comparison rule for the test is then different: instead of requiring all of a set of p individual values to fall within the prediction interval for the test to pass, only the average of the p future values should not fall outside the prediction limit (USEPA 2009).

6.2.1   Prediction Intervals for a Normal Distribution

Prediction Intervals for Future Observations

The guidance document *Statistical Analysis of Groundwater Monitoring Data at RCRA Facilities: Unified Guidance* (USEPA 2009) contains an example on page 18–9 where arsenic concentrations (ppb) are measured quarterly at a single well at a solid waste landfill. The first 3 years of the sample represent the background period and the fourth year is the compliance period. The data for this example are stored in EPA.09.Ex.18.1.arsenic.df in ENVSTATS.

```
> EPA.09.Ex.18.1.arsenic.df

  Year Sampling.Period Arsenic.ppb
1    1      Background        12.6
2    1      Background        30.8
3    1      Background        52.0
4    1      Background        28.1
5    2      Background        33.3
6    2      Background        44.0
7    2      Background         3.0
```

8	2	Background	12.8
9	3	Background	58.1
10	3	Background	12.6
11	3	Background	17.6
12	3	Background	25.3
13	4	Compliance	48.0
14	4	Compliance	30.3
15	4	Compliance	42.5
16	4	Compliance	15.0

Combining all of the observations from the background period and assuming these data come from a normal distribution, the exact one-sided upper 95 % prediction limit for the next $k = 4$ future observations is 72.9 ppb:

```
> attach(EPA.09.Ex.18.1.arsenic.df)

> predIntNorm(Arsenic.ppb[Sampling.Period == "Background"],
    k = 4, pi.type = "upper", conf.level = 0.95,
    method = "exact")

Results of Distribution Parameter Estimation
--------------------------------------------

Assumed Distribution:           Normal

Estimated Parameter(s):         mean = 27.51667
                                sd   = 17.10119

Estimation Method:              mvue

Data:
        Arsenic.ppb[Sampling.Period == "Background"]

Sample Size:                    12

Prediction Interval Method:     exact

Prediction Interval Type:       upper

Confidence Level:               95%

Number of Future Observations:  4

Prediction Interval:            LPL =      -Inf
                                UPL = 72.90375
```

and the one based on the Bonferroni method is 73.7 ppb:

```
> predIntNorm(Arsenic.ppb[Sampling.Period == "Background"],
    k = 4, pi.type = "upper",
    conf.level = 0.95)$interval$limits["UPL"]

     UPL
73.67237
```

The four observed values of arsenic in year 4 (the compliance period) are all below both of these prediction limits, so there is no evidence of contamination. Of course, even if one or more observations during the compliance period had exceeded the prediction limit, the evidence for "contamination" would depend on the assumption that "background" conditions had not changed.

Prediction Intervals for Future Means

For normally-distributed data, for the same background sample size and false positive rate, the power of the prediction limit for a future mean based on p future observations is generally higher than for a prediction limit for the next p individual future observations (USEPA 2009). Chapter 2 discussed functions in EnvStats for computing the power associated with a test based on prediction intervals. Figure 18–1 in USEPA (2009) demonstrates these power differences, and you can reproduce these figures in EnvStats using the accompanying scripts (see Chap. 1 for information on where these scripts are located).

USEPA (2009) contains an example on page 18–15 where chrysene concentrations (ppb) are measured at two background wells and one compliance well. The data for this example are stored in EPA.09.Ex.18.2.chrysene.df in ENVSTATS.

```
> EPA.09.Ex.18.2.chrysene.df
```

	Month	Well	Well.type	Chrysene.ppb
1	1	Well.1	Background	6.9
2	2	Well.1	Background	27.3
3	3	Well.1	Background	10.8
4	4	Well.1	Background	8.9
5	1	Well.2	Background	15.1
6	2	Well.2	Background	7.2
7	3	Well.2	Background	48.4
8	4	Well.2	Background	7.8
9	1	Well.3	Compliance	68.0
10	2	Well.3	Compliance	48.9
11	3	Well.3	Compliance	30.1
12	4	Well.3	Compliance	38.1

Combining the observations from the two background wells and assuming these data come from a lognormal distribution, the exact one-sided upper 99 % prediction limit for the mean of the next four future log-transformed observations is 3.85 log(ppb):

```
> attach(EPA.09.Ex.18.2.chrysene.df)

> predIntNorm(log(Chrysene.ppb)[Well.type == "Background"],
    n.mean = 4, k = 1, pi.type = "upper", conf.level = 0.99)

Results of Distribution Parameter Estimation
--------------------------------------------

Assumed Distribution:           Normal

Estimated Parameter(s):         mean = 2.5533006
                                sd   = 0.7060038

Estimation Method:              mvue

Data:
                    log(Chrysene.ppb)[Well.type == "Background"]

Sample Size:                    8

Prediction Interval Method:     exact

Prediction Interval Type:       upper

Confidence Level:               99%

Number of Future Averages:      1

Sample Size for Averages:       4

Prediction Interval:            LPL =      -Inf
                                UPL = 3.849427
```

(The exact method and the Bonferroni method are identical for one future observation or one future mean.) The mean of the log-transformed values at the compliance well is 3.79 log(ppb), so there is no evidence of contamination.

```
> mean(log(Chrysene.ppb)[Well.type == "Compliance"])

[1] 3.788506
```

6.2.2   *Prediction Intervals for a Lognormal Distribution*

A prediction interval for a lognormal distribution is constructed by simply taking the natural logarithm of the observations and constructing a prediction interval based on the normal distribution, then exponentiating the prediction limits to produce a prediction interval on the original scale of the data (Hahn and Meeker

1991, p. 73). In fact, you can use any monotonic transformation of the observations that you think induces normality (e.g., a Box-Cox power transformation), compute the prediction interval on the transformed scale, and then use the inverse transformation on the prediction limits to produce a prediction interval on the original scale. To construct a prediction interval for a lognormal distribution using ENVSTATS, type commands similar to those shown in the previous section for a normal distribution, except instead of using predIntNorm use the function predIntLnorm or predIntLnormAlt and the untransformed observations.

Prediction Intervals for Future Observations

We can use a prediction interval to compare the TcCB concentrations between the Cleanup and Reference areas (see Figs. 1.1, 1.2, and 1.3 in Chap. 1). Based on the data from the Background area, the one-sided upper 95 % prediction limit for the next $k = 77$ observations (there are 77 observations in the Cleanup area) is 2.68 ppb:

```
> attach(EPA.94b.tccb.df)

> predIntLnorm(TcCB[Area == "Reference"], k = 77,
    method = "exact", pi.type = "upper", conf.level = 0.95)

Results of Distribution Parameter Estimation
--------------------------------------------------

Assumed Distribution:              Lognormal

Estimated Parameter(s):            meanlog = -0.6195712
                                   sdlog   =  0.4679530

Estimation Method:                 mvue

Data:                              TcCB[Area == "Reference"]

Sample Size:                       47

Prediction Interval Method:        exact

Prediction Interval Type:          upper

Confidence Level:                  95%

Number of Future Observations:     77

Prediction Interval:               LPL = 0.000000
                                   UPL = 2.681076
```

There are seven observations in the Cleanup area larger than 2.68:

```
> sum(TcCB[Area=="Cleanup"] > 2.68)
```

```
[1] 7
```

so the prediction interval indicates residual contamination is present in the Cleanup area. Note that both Student's t-test and the Wilcoxon rank sum test do not yield a significant difference between the two areas.

Prediction Intervals for Future Geometric Means

Revisiting the example from Sect. 6.2.1, instead of log-transforming the chrysene data and using the function predIntNorm, we can use the original data and use predIntLnorm. The exact one-sided upper 99 % prediction limit for the *geometric* mean of the next four future observations is 47.0 ppb:

```
> attach(EPA.09.Ex.18.2.chrysene.df)   #if not attached
```

```
> predIntLnorm(Chrysene.ppb[Well.type == "Background"],
    n.geomean = 4, k = 1, pi.type = "upper",
    conf.level = 0.99)
```

```
Results of Distribution Parameter Estimation
--------------------------------------------

Assumed Distribution:            Lognormal

Estimated Parameter(s):          meanlog = 2.5533006
                                 sdlog   = 0.7060038

Estimation Method:               mvue

Data:
                       Chrysene.ppb[Well.type == "Background"]

Sample Size:                     8

Prediction Interval Method:      exact

Prediction Interval Type:        upper

Confidence Level:                99%

Number of Future
Geometric Means:                 1
```

```
Sample Size for
Geometric Means:                          4

Prediction Interval:                      LPL =   0.00000
                                          UPL = 46.96613
```

The geometric mean of the values at the compliance well is 44.2 ppb, so there is no evidence of contamination.

```
> geoMean(Chrysene.ppb[Well.type == "Compliance"])

[1] 44.19034
```

6.2.3   Prediction Intervals for a Gamma Distribution

Following the suggestion of Singh et al. (2002; 2010a, b), instead of assuming a lognormal distribution for the data in the examples of the previous section, we could instead assume the data follow a gamma distribution. Prediction intervals for a gamma distribution are constructed by using a power transformation to approximate a normal distribution, computing the prediction interval for a normal distribution based on the transformed data, then transforming the prediction interval back to the original scale. Choices for what power transformation to use to approximate normality include the power transformation of Kulkarni and Powar (2010), the cube-root transformation (Wilson and Hilferty 1931; Krishnamoorthy et al. 2008), and the fourth-root transformation (Hawkins and Wixley 1986). See Millard et al. (2014) or the help file for predIntGamma for details

Prediction Intervals for Future Observations

For the first example that involves comparing the TcCB concentrations between the Cleanup and Reference areas, based on the data from the Background area, the one-sided upper 95 % prediction limit for the next $k = 77$ observations is 2.14 ppb as compared to 2.68 ppb computed using a lognormal assumption:

```
> predIntGamma(TcCB[Area == "Reference"], k = 77,
     method = "exact", pi.type = "upper", conf.level = 0.95)

Results of Distribution Parameter Estimation
--------------------------------------------------


Assumed Distribution:             Gamma

Estimated Parameter(s):           shape = 4.8659316
                                  scale = 0.1230002

Estimation Method:                mle

Data:                             TcCB[Area == "Reference"]
```

```
Sample Size:                    47

Prediction Interval Method:     exact using
                                Kulkarni & Powar (2010)
                                transformation to Normality
                                based on mle of 'shape'

Normal Transform Power:         0.246

Prediction Interval Type:       upper

Confidence Level:               95%
Number of Future Observations:  77

Prediction Interval:            LPL = 0.000000
                                UPL = 2.143873
```

Prediction Intervals for Future Transformed Means

For the second example involving chrysene concentrations (ppb) measured at two background wells and one compliance well, combining the observations from the two background wells and assuming these data come from a gamma distribution, the exact one-sided upper 99 % prediction limit for the **transformed** mean of the next four future observations is 45 ppb:

```
> predInt.list <- predIntGamma(
    Chrysene.ppb[Well.type == "Background"], n.transmean = 4,
    k = 1, pi.type = "upper", conf.level = 0.99)

> predInt.list

Results of Distribution Parameter Estimation
--------------------------------------------


Assumed Distribution:           Gamma

Estimated Parameter(s):         shape = 2.127279
                                scale = 7.779891

Estimation Method:              mle

Data:
      Chrysene.ppb[Well.type == "Background"]

Sample Size:                    8
```

```
Prediction Interval Method:        exact using
                                   Kulkarni & Powar (2010)
                                   transformation to Normality
                                   based on mle of 'shape'

Normal Transform Power:            0.246

Prediction Interval Type:          upper

Confidence Level:                  99%

Number of Future
Transformed Means:                 1

Sample Size for
Transformed Means:                 4

Prediction Interval:               LPL =   0.00000
                                   UPL = 45.02989
```

Recall that in Sect. 6.2.2 when we assumed a lognormal distribution, we computed the **geometric** mean for the compliance well; that is, we computed the mean based on the log-transformed data and then transformed that mean back to the original scale using the exponential function. Similarly, in this example we need to transform the original data from the compliance well using the transformation that was used to construct the prediction interval assuming a gamma distribution, compute the mean based on these transformed data, then back-transform this mean to the original scale.

```
> trans.power <- predInt.list$interval$normal.transform.power

> trans.power

[1] 0.246

> mean.of.trans <- mean(
    Chrysene.ppb[Well.type == "Compliance"] ^ trans.power)

> mean.of.trans ^ (1 / trans.power)

[1] 44.69182
```

Since 44.7 is less than the UPL of 45 ppb (just barely!), there is no evidence of contamination. **However,** we will see in Chap. 7 that the goodness-of-fit test rejects the hypothesis that the chrysene data at the background wells comes from a gamma distribution, so **the assumption of a gamma distribution is probably not valid**.

6.2.4   Nonparametric Prediction Intervals

You can construct a prediction interval without making any assumption about the distribution of the background data, except that the distribution is continuous. This kind of prediction intervals is called a **nonparametric prediction interval**, and it is based on the ranked data. Usually the prediction interval is based on the maximum and/or the minimum of the background data, but it can be based on any order statistics you choose. Of course, a nonparametric prediction interval still requires the assumption that the distribution of future observations is the same as the distribution of the observations used to create the prediction interval. See Millard et al. (2014) for a detailed discussion.

Nonparametric Prediction Intervals for Future Observations

Table 6.2 illustrates the confidence levels associated with a one-sided upper prediction interval for the next $m = 3$ observations, based on various sample sizes, assuming the upper prediction limit is the maximum value.

Sample size (n)	Confidence level (%)
5	62
10	77
15	83
20	87
25	89
50	94
75	96
100	97

Table 6.2 Confidence levels for one-sided upper nonparametric prediction interval for the next $m = 3$ observations, based on using the maximum value as the upper prediction limit

The values for this table were created with the following commands:

```
> n <- c(seq(5, 20, by = 5), seq(25, 100, by = 25))
> round(100 *
    predIntNparConfLevel(n = n, m = 3, pi.type = "upper"))
```

You can see that a confidence level greater than 95 % cannot be achieved until the sample size is larger than $n = 50$.

Month	Background			Compliance
	Well 1	Well 2	Well 3	Well 4
1	<5	7	<5	
2	<5	6.5	<5	
3	8	<5	10.5	7.5
4	<5	6	<5	<5
5	9	12	<5	8
6	10	<5	9	14

Table 6.3 Trichloroethylene data (ppb) from groundwater monitoring wells

USEPA (2009, pp.18–19) gives an example of constructing a nonparametric prediction interval for the next $m = 4$ monthly observations of trichloroethylene concentrations (ppb) in groundwater at a downgradient well, based on observations from three background wells. These data are shown in Table 6.3 and stored in the data frame EPA.09.Ex.18.3.TCE.df in ENVSTATS.

```
> EPA.09.Ex.18.3.TCE.df

    Month Well  Well.type TCE.ppb.orig TCE.ppb Censored
1       1 BW-1 Background           <5     5.0     TRUE
2       2 BW-1 Background           <5     5.0     TRUE
...
23      5 CW-4 Compliance            8     8.0    FALSE
24      6 CW-4 Compliance           14    14.0    FALSE
```

The three background wells were sampled once per month for 6 months. The compliance well was only sampled in months 3–6. The EPA guidance document combines all of the observations from the three background wells ($n = 18$) and uses the maximum value 12 as an upper prediction limit for the next $m = 4$ observations at the compliance well. This produces an 82 % upper prediction interval.

```
> with(EPA.09.Ex.18.3.TCE.df,
    predIntNpar(TCE.ppb[Well.type == "Background"], m = 4,
    lb = 0, pi.type = "upper"))

Results of Distribution Parameter Estimation
--------------------------------------------------

Assumed Distribution:            None

Data:
                        TCE.ppb[Well.type == "Background"]

Sample Size:                     18

Prediction Interval Method:      Exact

Prediction Interval Type:        upper

Confidence Level:                82%

Prediction Limit Rank(s):        18

Number of Future Observations:   4

Prediction Interval:             LPL =   0
                                 UPL = 12
```

Since one of the values from the compliance well lies above the upper prediction limit, we might conclude there is evidence of contamination at the compliance well, but we should keep in mind that given the way we constructed our prediction interval, we would incorrectly declare contamination present when in fact it is not present about 18 % (100–82 %) of the time. As USEPA (2009, pp. 18–19) states: "Only additional background data and/or use of a retesting strategy would lower the false positive rate."

Nonparametric Prediction Intervals for a Single Future Median

Constructing a prediction interval for a future median of order 3 (i.e., three future observations will be used to construct the median) is equivalent to constructing a simultaneous prediction interval for the next 2 of 3 observations (see the Sect. 6.3). In general, a prediction interval for a future median of order m is equivalent to a prediction interval for the next $(m + 1)/2$ of m observations as long as m is an odd number.

| Month | Background | | | Compliance |
	Well 1	Well 2	Well 3	Well 4
1	<5	9.2	<5	
2	<5	<5	5.4	
3	7.5	<5	6.7	
4	<5	6.1	<5	
5	<5	8.0	<5	
6	<5	5.9	<5	<5
7	6.4	<5	<5	7.8
8	6.0	<5	<5	10.4

Table 6.4 Xylene data (ppb) from groundwater monitoring wells

USEPA (2009, pp.18–21) gives an example of constructing a nonparametric prediction interval for a future median of order 3 (i.e., three future observations will be used to construct the median) using monthly observations of xylene concentrations (ppb) in groundwater at a downgradient well, based on observations from three background wells. These data are stored in the data frame EPA.09.Ex.18.4.xylene.df in ENVSTATS and shown in Table 6.4.

```
> EPA.09.Ex.18.4.xylene.df

   Month   Well   Well.type Xylene.ppb.orig Xylene.ppb Censored
1      1 Well.1  Background             <5        5.0     TRUE
2      2 Well.1  Background             <5        5.0     TRUE
...
31     7 Well.4  Compliance            7.8        7.8    FALSE
32     8 Well.4  Compliance           10.4       10.4    FALSE
```

Combining all of the observations from the three background wells ($n = 24$) and using the maximum value 9.2 ppb as an upper prediction limit produces a 99 % upper prediction interval for a future median of order 3.

```
> with(EPA.09.Ex.18.4.xylene.df,
    predIntNpar(Xylene.ppb[Well.type == "Background"], k = 2,
    m = 3, lb = 0, pi.type = "upper")
```

Results of Distribution Parameter Estimation

Assumed Distribution: None

Data:
                        Xylene.ppb[Well.type == "Background"]

Sample Size: 24

Prediction Interval Method: Exact

Prediction Interval Type: upper

Confidence Level: 99%

Prediction Limit Rank(s):   24

Minimum Number of
Future Observations
Interval Should Contain: 2

Total Number of
Future Observations: 3

Prediction Interval: LPL = 0.0
                                 UPL = 9.2

Since the median of the values at the compliance well is 7.8 ppb and therefore less than the upper prediction limit, there is no evidence of contamination at the compliance well, in spite of the fact that the maximum value at the compliance well is greater than the upper prediction limit.

6.3   Simultaneous Prediction Intervals

Analyzing data from a groundwater monitoring program involves several difficulties, including trying to control for natural spatial and temporal variability, and sometimes dealing with nondetect values. One of the main statistical problems that plague groundwater monitoring programs at hazardous and solid waste facilities is the requirement of testing several wells and several constituents at each well on each sampling occasion. The number of constituents monitored can range from around 5 to 60 or more, and some facilities may have as many as 150 monitoring wells (Davis and McNichols 1999). This is an obvious multiple comparisons problem, and the naive approach of using a prediction interval with a conventional confidence level (e.g., 95 % or 99 %) for each comparison of a compliance well with background for each chemical of concern leads to a very high probability of at least one declaration of contamination on each sampling occasion, when in fact no contamination has occurred at any of the wells at any time for any of the chemicals of concern. This problem was pointed out several years ago by Millard (1987a) and others.

Davis and McNichols (1987, 1994b, 1999) proposed simultaneous prediction intervals as a way of controlling the site-wide false positive rate (SWFPR) while maintaining adequate power to detect contamination in the groundwater. A *simultaneous prediction interval* with confidence level $(1-\alpha)100$ % is a prediction interval that will contain a specified number of future observations with probability $(1-\alpha)100$ % for each of r future sampling occasions, where r is some pre-specified positive integer. The quantity r may actually refer to r distinct future sampling occasions in time, r distinct compliance wells sampled on one future sampling occasion, or the product of the number of future sampling occasions and number of wells. In any of these cases, it is assumed that the distribution of concentrations is constant over all r "future sampling occasions."

There are several ways to define a rule for a simultaneous prediction interval. EnvStats includes functions for the following three rules:

- **The k-of-m Rule**. For the k-of-m rule, at least k of the next m future observations will fall in the prediction interval with probability $(1-\alpha)100$ % on each of the r future sampling occasions. If observations are being taken sequentially, for a particular sampling occasion (or monitoring well), up to m observations may be taken, but once k of the observations fall within the prediction interval, sampling can stop. If $m-(k-1)$ observations fall outside the prediction interval, then contamination is declared to be present. For example, suppose we have $r = 5$ monitoring wells and we want to use the 1-of-3 rule (i.e., $k = 1$ and $m = 3$). Then for the ith monitoring well ($i = 1, 2, 3, 4, 5$), if the first observation is in the interval, we can stop. If the first observation is outside the interval, we have to wait a specified time (e.g., a few weeks), and take a second observation. If the second observation is in the interval, we can stop. If the second observation is outside the interval, then we have to wait a specified time and take a third observation. If the third observation is in the interval, we can stop. If the third observation is outside the interval, then contamination is declared to be present. (Note that in the case $k = m$ and $r = 1$, a simultaneous prediction interval reduces to the simple prediction interval we have already discussed in Sect. 6.2.)

- **California Rule**. For the California rule, with probability $(1-\alpha)100$ %, for each of the r future sampling occasions, either the first observation will fall in the prediction interval, or else all of the next $m-1$ observations will fall in the prediction interval. That is, if the first observation falls in the prediction interval then sampling can stop. Otherwise, up to $m-1$ more observations must be taken (with a sufficient waiting time between sampling occasions). If any of these subsequent $m-1$ observations falls outside the interval, we declare contamination is present.

- **Modified California Rule**. For the Modified California rule, with probability $(1-\alpha)100$ %, for each of the r future sampling occasions, either the first observation will fall in the prediction interval, or else at least 2 out of the next 3 observations will fall in the prediction interval. That is, if the first observation falls in the prediction interval then sampling can stop. Otherwise, up to 3 more observations must be taken (with a sufficient waiting time between sampling occasions). If any two of these next three observations fall into the interval then sampling can stop. Otherwise, contamination is declared to be present.

Just as in the case of regular prediction intervals, instead of constructing intervals for future *observations*, it is possible to construct simultaneous prediction intervals for the *mean* or *median* of future observations.

Although simultaneous prediction intervals help us control the Type I error rate (the probability of declaring contamination when it is not present) over r future sampling occasions (or monitoring wells), we need to control the Type I error rate over all future sampling occasions, all monitoring wells, and all constituents (chemicals and physical properties) we monitor. USEPA (2009, Chap. 19) gives guidelines for setting the confidence level in order to control the *annual* SWFPR (α), and suggests setting the annual SWFPR to 10 % (USEPA 2009, pp. 6–4).

For parametric simultaneous prediction intervals, USEPA (2009) suggests using a confidence level based on adjusting for the number of well-constituent pairs, i.e., the number of monitoring wells (n_w) times the number of constituents (n_c):

$$Confidence\ Level = \left(1-\alpha\right)^{1/(n_w n_c)}\ 100\% \qquad (6.4)$$

In this case, the number of future observations r is set to the number of evaluations per year (n_E), so for annual evaluations $r=1$, semi-annual evaluations $r=2$, etc.

For nonparametric simultaneous prediction intervals, for performing *interwell* tests in which all monitoring wells are compared to the same background data, USEPA (2009) suggests using a confidence level based on adjusting for the number of constituents:

$$Confidence\ Level = \left(1-\alpha\right)^{1/n_c}\ 100\% \qquad (6.5)$$

and setting the number of future sampling occasions r to the product of the number of wells (n_w) and the number of evaluations per year (n_E). For performing *intrawell* tests in which each monitoring well is compared to its own background data, USEPA (2009) suggests using the confidence level defined in Eq. 6.4 above.

6.3.1   Simultaneous Prediction Intervals for a Normal Distribution

Using the background period arsenic data introduced in Sect. 6.2.1 and assuming there are $n_w = 20$ compliance wells to be monitored semi-annually (i.e., $r = 2$), a total of $n_c = 10$ constituents (including arsenic), and negligible spatial variability so that you can use interwell testing, we can construct 90 % upper simultaneous prediction limits based on various rules. Here we will consider the 1-of-2 rule, the 1-of-3 rule, the Modified California rule, and the 1-of-2 rule based on means of order 2. Using Eq. 6.4, the confidence level is set to 99.94733%:

```
> nw <- 20

> nc <- 10

> conf.level <- (1 - 0.1)^(1 / (nc * nw))

> conf.level

[1] 0.9994733
```

Now use the background period arsenic data to construct the upper prediction limit for each of the rules. For the 1-of-2 rule, the upper limit is 80.1 ppb:

```
> attach(EPA.09.Ex.18.1.arsenic.df)   #If not already attached

> As.Bkgrd <- Arsenic.ppb[Sampling.Period == "Background"]

> predIntNormSimultaneous(As.Bkgrd, k = 1, m = 2, r = 2,
    rule = "k.of.m", pi.type = "upper",
    conf.level = conf.level)

Results of Distribution Parameter Estimation
--------------------------------------------


Assumed Distribution:            Normal

Estimated Parameter(s):          mean = 27.51667
                                 sd   = 17.10119

Estimation Method:               mvue

Data:                            As.Bkgrd

Sample Size:                     12

Prediction Interval Method:      exact

Prediction Interval Type:        upper
```

```
Confidence Level:                   99.94733%

Minimum Number of
Future Observations
Interval Should Contain
(per Sampling Occasion):            1

Total Number of
Future Observations
(per Sampling Occasion):            2

Number of Future
Sampling Occasions:                 2

Prediction Interval:                LPL =      -Inf
                                    UPL = 80.09079
```

For the 1-of-3 rule the limit is 65.3 ppb:

```
> predIntNormSimultaneous (As.Bkgrd, k = 1, m = 3, r = 2,
    rule = "k.of.m", pi.type = "upper",
    conf.level = conf.level)$interval$limits["UPL"]

    UPL
65.29204
```

for the Modified California rule the limit is 71.1 ppb:

```
> predIntNormSimultaneous (As.Bkgrd, r = 2,
    rule = "Modified.CA", pi.type = "upper",
    conf.level = conf.level)$interval$limits["UPL"]

    UPL
71.11351
```

and for the 1-of-2 rule using means of order 2 the limit is 67.5 ppb:

```
> predIntNormSimultaneous (As.Bkgrd, n.mean = 2, k = 1,
    m = 2, r = 2, rule = "k.of.m", pi.type = "upper",
    conf.level = conf.level)$interval$limits["UPL"]

    UPL
67.54322
```

Using the following commands, we can construct a data frame showing the upper prediction limits for each of the rules, along with the power of detecting a change in concentration of three standard deviations at any of the 20 compliance wells during the course of a year, as well as the total number of potential samples that may have to be taken.

```
> n <- sum (!is.na (As.Bkgrd))

> rule.vec <- c("k.of.m", "k.of.m", "Modified.CA", "k.of.m")

> n.mean.vec <- c(1, 1, 1, 2)

> m.vec <- c(2, 3, 4, 2)

> n.rules <- length(rule.vec)

> UPL.vec <- rep(as.numeric(NA), n.rules)

> for(i in 1:n.rules)
    UPL.vec[i] <- predIntNormSimultaneous (As.Bkgrd,
    n.mean = n.mean.vec[i], k = 1, m = m.vec[i], r = 2,
    rule = rule.vec[i], pi.type = "upper",
    conf.level = conf.level)$interval$limits["UPL"]

> Power.vec <- predIntNormSimultaneousTestPower (n = n,
    k = 1, m = m.vec, n.mean = n.mean.vec, r = 2,
    rule = rule.vec, delta.over.sigma = 3,
    pi.type = "upper", conf.level = conf.level)

> data.frame(Rule = rule.vec, k = rep(1, n.rules),
    m = m.vec, N.Mean = n.mean.vec, UPL = round(UPL.vec, 1),
    Power = round(Power.vec, 2),
    Total.Samples = n.mean.vec * m.vec * r)
```

	Rule	k	m	N.Mean	UPL	Power	Total.Samples
1	k.of.m	1	2	1	80.1	0.46	4
2	k.of.m	1	3	1	65.3	0.70	6
3	Modified.CA	1	4	1	71.1	0.70	8
4	k.of.m	1	2	2	67.5	0.81	8

We can see that the 1-of-2 rule using means of order 2 gives the highest power (81 %) for detecting a change in concentration of three standard deviations at any of the 20 compliance wells during the course of a year, but it may potentially involve taking up to eight samples during the course of the year, which might not be feasible either in terms of avoiding temporal correlation or in terms of the time and cost involved to collect so many samples. On the other hand, although the 1-of-2 rule for future single observations requires the least number of potential samples, it has poor power (46 %).

6.3.2 Simultaneous Prediction Intervals for a Lognormal Distribution

Just as for a standard prediction interval for a lognormal distribution, a simultaneous prediction interval for a lognormal distribution is constructed by simply taking the natural logarithm of the observations and constructing a simultaneous prediction interval based on the normal distribution, then exponentiating the prediction limits to produce a simultaneous prediction interval on the original scale of the data. To construct a simultaneous prediction interval for a lognormal distribution you can use the ENVSTATS functions `predIntLnormSimultaneous` or `predIntLnormAltSimultaneous`.

USEPA (2009) contains an example on page 19–17 in which sulfate concentrations (mg/l) are to be monitored at $n_w = 50$ compliance wells on a semi-annual basis. There are $n_c = 10$ constituents total (including sulfate), and negligible spatial variability so that you can use interwell testing. Because the regulating authority will only allow up to two resamples per exceedence of the background concentration limit, we cannot consider a 1-of-4 or Modified California rule. Here we will look at the 1-of-2 and 1-of-3 plan. The $n = 25$ background sulfate observations are stored in the data frame `EPA.09.Ex.19.1.sulfate.df` in EnvStats:

```
> EPA.09.Ex.19.1.sulfate.df[ , -(2:4)]

      Well       Date Sulfate.mg.per.l log.Sulfate.mg.per.l
1   GW-01 1999-07-08            63.0            4.143135
2   GW-01 1999-09-12            51.0            3.931826
3   GW-01 1999-10-16            60.0            4.094345
...
23  GW-09 2000-10-24            85.5            4.448516
24  GW-09 2002-12-01           188.0            5.236442
25  GW-09 2003-03-24           150.0            5.010635
```

A check for normality of the pooled background sulfate measurements indicates a log transformation is appropriate. Using Eq. 6.4, the confidence level is set to 99.97893 %:

```
> nw <- 50

> nc <- 10

> conf.level <- (1 - 0.1)^(1 / (nc * nw))

> conf.level

[1] 0.9997893
```

We can compare the power of detecting a change in concentration of three standard deviations (on the log scale) at any of the 50 compliance wells during the course of a year for the 1-of-2 rule versus the 1-of-3 rule:

```
> predIntNormSimultaneousTestPower(n = 25, k = 1, m = 2:3,
    r = 2, rule = "k.of.m", delta.over.sigma = 3,
    pi.type = "upper", conf.level = conf.level)

[1] 0.5776416 0.8023368
```

Since the power of the 1-of-3 test is much better than the 1-of-2 test, we will compute the upper prediction limit based on the 1-of-3 test.

```
> with(EPA.09.Ex.19.1.sulfate.df,
    predIntLnormSimultaneous(Sulfate.mg.per.l,
    k = 1, m = 3, r = 2, rule = "k.of.m", pi.type = "upper",
    conf.level = conf.level))
```

```
Results of Distribution Parameter Estimation
--------------------------------------------
```

Assumed Distribution: Lognormal

Estimated Parameter(s):   meanlog = 4.3156194
                                   sdlog   = 0.3756697

Estimation Method: mvue

Data: Sulfate.mg.per.l

Sample Size: 25

Prediction Interval Method: exact

Prediction Interval Type: upper

Confidence Level: 99.97893%

Minimum Number of
Future Observations
Interval Should Contain
(per Sampling Occasion):   1

Total Number of
Future Observations
(per Sampling Occasion):   3

Number of Future
Sampling Occasions: 2

Prediction Interval: LPL =   0.0000
                                   UPL = 159.5497

So for each of the semi-annual sampling occasions and each of the 50 compliance wells, if a sulfate concentration is greater than 159.5 mg/l for the first sample, then you need to re-sample and compare again to 159.5 mg/l. If the first re-sample is above this limit you need to take a second re-sample. If the second re-sample is also above this limit, then you can declare contamination.

6.3.3 Simultaneous Prediction Intervals for a Gamma Distribution

Again, following the suggestion of Singh et al. (2002, 2010a, b), instead of assuming a lognormal distribution for the data in the example of the previous section, we could instead assume the data follow a gamma distribution. A goodness-of-fit test indicates the pooled background sulfate measurements fit a gamma distribution. The upper prediction limit based on the 1-of-3 test is 153.2 mg/l as compared to 159.5 g/l assuming a lognormal distribution:

```
> with(EPA.09.Ex.19.1.sulfate.df,
    predIntGammaSimultaneous(Sulfate.mg.per.l,
    k = 1, m = 3, r = 2, rule = "k.of.m", pi.type = "upper",
    conf.level = conf.level)$interval$limits["UPL"]

     UPL
153.3232
```

6.3.4 Simultaneous Nonparametric Prediction Intervals

Chou and Owen (1986) developed the theory for nonparametric simultaneous prediction limits for various rules, including the 1-of-m rule. Their theory, however, does not cover the California or Modified California rules, and uses an r-fold summation involving a minimum of 2^r terms. Davis and McNichols (1994b, 1999) extended the results of Chou and Owen (1986) to include the California and Modified California rule, and developed algorithms that involve summing far fewer terms.

Like a standard nonparametric prediction interval, a simultaneous nonparametric prediction interval is based on the order statistics from the sample. For a one-sided upper simultaneous nonparametric prediction interval, the upper prediction limit is usually the largest observation in the background data, but it could be the next largest or any other order statistic. Similarly, for a one-sided lower simultaneous nonparametric prediction interval, the lower prediction limit is usually the smallest observation. Simultaneous nonparametric prediction intervals can also be extended to the case of predicting future medians instead of future observations.

Event	BG-1	BG-2	BG-3	BG-4	CW-1	CW-2
1	0.21	<0.2	<0.2	<0.2	0.22	0.36
2	<0.2	<0.2	0.23	0.25	0.20	0.41
3	<0.2	<0.2	<0.2	0.28	<0.2	0.28
4	<0.2	0.21	0.23	<0.2	0.25	0.45
5	<0.2	<0.2	0.24	<0.2	0.24	0.43
6					<0.2	0.54

Table 6.5 Mercury data (ppb) from groundwater monitoring wells

USEPA (2009) contains an example on page 19–33 in which mercury concentrations (ppb) are to be monitored at $n_w = 10$ compliance wells on an annual basis.

There are $n_c = 5$ constituents total (including mercury), and negligible spatial variability so that you can use interwell testing. Table 6.5 shows the $n = 20$ background observations (collected from four different wells), along with data from two of the 10 compliance wells. These data are stored in the data frame EPA.09.Ex.19.5.mercury.df in ENVSTATS.

```
> EPA.09.Ex.19.5.mercury.df

   Event Well  Well.type Mercury.ppb.orig Mercury.ppb Censored
1      1 BG-1 Background             0.21        0.21    FALSE
2      2 BG-1 Background             <.2         0.20     TRUE
...
35     5 CW-2 Compliance            0.43        0.43    FALSE
36     6 CW-2 Compliance            0.54        0.54    FALSE
```

Because there are so many non-detect values, we need to use a nonparametric approach. Using Eq. 6.5, the confidence level is set to 97.91484 %, and the corresponding per-test Type I error rate is 2.085164 %:

```
> nc <- 5

> conf.level <- (1 - 0.1)^(1 / nc)

> conf.level

[1] 0.9791484

> alpha <- 1 - conf.level

> alpha

[1] 0.02085164
```

The number of future sampling occasions r is set to the product of the number of compliance wells and the number of evaluations per year:

```
> nw <- 10

> ne <- 1

> r <- nw * ne
```

Now we need to determine which sampling plans will yield a per-test Type I error less than or equal to the required 2.1 % level. Here we will consider six candidate rules:

1) 1-of-2
2) 1-of-3
3) 1-of-4
4) Modified California
5) 1-of-1 for the median of 3 future values. (This plan is equivalent to the 2-of-3 plan for single observations.)
6) 1-of-2 for the median of 3 future values.

and we will consider using the maximum, second largest, or third largest value of the background data as the upper simultaneous prediction limit (i.e., a total of 18 candidate sampling plans). First we will create some data objects that store information about the six different rules:

```
> rule.vec <- c(rep("k.of.m", 3), "Modified.CA",
    rep("k.of.m", 2))

> k.vec <- rep(1, 6)

> m.vec <- c(2:4, 4, 1, 2)

> n.median.vec <- c(rep(1, 4), rep(3, 2))

> n.plans <- length(rule.vec)
```

Next we'll compute the per-test Type I error associated with using the maximum value of the background data as the upper simultaneous prediction limit:

```
> n <- 20

> alpha.vec.Max <- 1 - predIntNparSimultaneousConfLevel(
    n = n, n.median = n.median.vec, k = k.vec, m = m.vec,
    r = r, rule = rule.vec, pi.type = "upper")
```

Next we'll compute the per-test Type I error associated with using the second largest and third largest value of the background data as the upper simultaneous prediction limit. The code for this looks just like the code above, except that we set the argument n.plus.one.minus.upl.rank equal to 2 or 3:

```
> alpha.vec.2nd <- 1 - predIntNparSimultaneousConfLevel(
    n = n, n.median = n.median.vec, k = k.vec, m = m.vec,
    r = r, rule = rule.vec, pi.type = "upper",
    n.plus.one.minus.upl.rank = 2)

> alpha.vec.3rd <- 1 - predIntNparSimultaneousConfLevel(
    n = n, n.median = n.median.vec, k = k.vec, m = m.vec,
    r = r, rule = rule.vec, pi.type = "upper",
    n.plus.one.minus.upl.rank = 3)
```

Now create a data frame listing all 18 of the plans, their associated per-test Type I error rate, and their associated upper prediction limit:

```
> attach(EPA.09.Ex.19.5.mercury.df)

> Bkgd.Hg.Sorted <- sort(Mercury.ppb[
    Well.type == "Background"], decreasing = TRUE)
```

```
> Candidate.Plans.df <- data.frame(Rule = rep(rule.vec, 3), k
    = rep(k.vec, 3), m = rep(m.vec, 3),
  Median.n = rep(n.median.vec, 3), Order.Statistic =
  rep(c("Max", "2nd", "3rd"), each = n.plans),
  Achieved.alpha = round(c(alpha.vec.Max, alpha.vec.2nd,
                           alpha.vec.3rd), 4),
  BG.Limit = rep(Bkgd.Hg.Sorted[1:3], each = n.plans))

> Candidate.Plans.df
```

	Rule	k	m	Median.n	Order.Statistic	Achieved.alpha	BG.Limit
1	k.of.m	1	2	1	Max	0.0395	0.28
2	k.of.m	1	3	1	Max	0.0055	0.28
3	k.of.m	1	4	1	Max	0.0009	0.28
4	Modified.CA	1	4	1	Max	0.0140	0.28
5	k.of.m	1	1	3	Max	0.0961	0.28
6	k.of.m	1	2	3	Max	0.0060	0.28
7	k.of.m	1	2	1	2nd	0.1118	0.25
8	k.of.m	1	3	1	2nd	0.0213	0.25
9	k.of.m	1	4	1	2nd	0.0046	0.25
10	Modified.CA	1	4	1	2nd	0.0516	0.25
11	k.of.m	1	1	3	2nd	0.2474	0.25
12	k.of.m	1	2	3	2nd	0.0268	0.25
13	k.of.m	1	2	1	3rd	0.2082	0.24
14	k.of.m	1	3	1	3rd	0.0516	0.24
15	k.of.m	1	4	1	3rd	0.0135	0.24
16	Modified.CA	1	4	1	3rd	0.1170	0.24
17	k.of.m	1	1	3	3rd	0.4166	0.24
18	k.of.m	1	2	3	3rd	0.0709	0.2

Eliminate plans that do not achieve the required per-test Type I error rate of 2.1 %:

```
> index <- Candidate.Plans.df$Achieved.alpha <= alpha

> Candidate.Plans.df <- Candidate.Plans.df[index, ]

> Candidate.Plans.df
```

	Rule	k	m	Median.n	Order.Statistic	Achieved.alpha	BG.Limit
2	k.of.m	1	3	1	Max	0.0055	0.28
3	k.of.m	1	4	1	Max	0.0009	0.28
4	Modified.CA	1	4	1	Max	0.0140	0.28
6	k.of.m	1	2	3	Max	0.0060	0.28
9	k.of.m	1	4	1	2nd	0.0046	0.25
15	k.of.m	1	4	1	3rd	0.0135	0.24

For the plans based on predicting individual observations, the ones that achieve the required per-test Type I error level are the 1-of-3 and the Modified California using the maximum background value, and the 1-of-4 using the maximum, second largest, or third largest background value. For plans based on predicting medians, only the 1-of-2 plan using the maximum background level meets the required per-test Type I error rate.

Looking at the six final candidate plans above and comparing them to the data for the two compliance wells in Table 6.5, we see that the first compliance well passes for each of the six plans since the first observed value is 0.22 ppb which is less than the upper simultaneous prediction limit for all of the plans. The second compliance well only passes for the first two plans and fails the last four.

One step that we did not perform yet was to look at the power of each plan (something that normally is done prior to choosing a specific plan to use and prior to actually comparing compliance well data to the upper prediction limit). In order to compute power, you need to make an assumption about the distribution of the background data. Here we will assume a normal distribution and compute the power of detecting a change in concentration of three standard deviations at any of the 10 compliance wells during the course of a year, as well as the total number of potential samples that may have to be taken.

```
> Power.vec <- predIntNparSimultaneousTestPower(n = n,
    n.median = Candidate.Plans.df[, "Median.n"],
    k = Candidate.Plans.df[, "k"],
    m = Candidate.Plans.df[, "m"], r = r,
    rule = as.character(Candidate.Plans.df[, "Rule"]),
    n.plus.one.minus.upl.rank = match(Candidate.Plans.df[,
        "Order.Statistic"] , c("Max", "2nd", "3rd")),
    delta.over.sigma = 3, pi.type = "upper", r.shifted = 1,
    distribution = "norm", method = "approx")

> data.frame(Candidate.Plans.df[, c("Rule", "k", "m",
    "Median.n", "Order.Statistic")],
    Power = round(Power.vec, 2), Total.Samples =
    Candidate.Plans.df$Median.n * Candidate.Plans.df$m * ne)
```

	Rule	k	m	Median.n	Order.Statistic	Power	Total.Samples
2	k.of.m	1	3	1	Max	0.65	3
3	k.of.m	1	4	1	Max	0.58	4
4	Modified.CA	1	4	1	Max	0.81	4
6	k.of.m	1	2	3	Max	0.91	6
9	k.of.m	1	4	1	2nd	0.78	4
15	k.of.m	1	4	1	3rd	0.87	4

The first two plans (under which the second compliance well passed) have the lowest power. The plan with the highest power, the 1-of-2 for medians of order 3, also requires the most potential resampling.

6.4 Tolerance Intervals

A *tolerance interval* for some population is an interval on the real line constructed so as to contain $\beta 100$ % of the population (i.e., $\beta 100$ % of all future observations), where $0 < \beta < 1$ (usually β is bigger than 0.5). The quantity $\beta 100$ % is called the *coverage*. (Note: Do not confuse our use of the symbol β here with the probability of a Type II error. The symbol β is used here to be consistent with previous literature on tolerance intervals.) Table 6.6 lists the functions available in ENVSTATS for constructing tolerance intervals.

Distribution	Function name	Description
Gamma	tolIntGamma	Construct a tolerance interval for a gamma
	tolIntGammaAlt	distribution
Normal	tolIntNorm	Construct a tolerance interval for a normal
		distribution
	tolIntNormK	Compute the value of K for a tolerance interval for a
		normal distribution
Lognormal	tolIntLnorm	Construct a tolerance interval for a lognormal distri-
	tolIntLnormAlt	bution
Poisson	tolIntPois	Construct a tolerance interval for a Poisson distribu-
		tion
Nonparametric	tolIntNpar	Construct a nonparametric tolerance interval

Table 6.6 Functions in ENVSTATS for constructing tolerance intervals

As with a prediction interval, the basic idea of a tolerance interval is to assume a particular probability distribution (e.g., normal, lognormal, etc.) for some process generating the data (e.g., quarterly observations of chemical concentrations in groundwater), compute sample statistics from a baseline sample, and then use these sample statistics to construct a tolerance interval, assuming the distribution of the data does not change in the future. For example, if X denotes a random variable from some population, and we know what the population looks like (e.g., N(10, 2)) so we can compute the quantiles of the population, then a $\beta 100$ % two-sided tolerance interval is given by:

$$\left[x_{1-\beta/2} \, , \, x_{\beta/2} \right] \tag{6.6}$$

where x_p denotes the pth quantile of the distribution of X. Similarly, a $\beta 100$ % one-sided upper tolerance interval is given by:

$$\left[-\infty \, , \, x_\beta \right] \tag{6.7}$$

and a $\beta 100$ % one-sided lower tolerance interval is given by:

$$\left[x_{1-\beta} \, , \, \infty \right] \tag{6.8}$$

Note that in the case when the distribution of X is known, a $\beta 100$ % tolerance interval is exactly the same as a $(1-\alpha)100$ % prediction interval for $k = 1$ future observation, where $\beta = 1-\alpha$ (see Eq. 6.1, 6.2, and 6.3).

Usually the true distribution of X is unknown, so the values of the tolerance limits have to be estimated based on estimating the parameters of the distribution of X. In this case, a tolerance interval is a random interval; that is, the lower and/or upper bounds are random variables computed based on sample statistics in the baseline sample. Given this uncertainty in the bounds, there are two ways to construct tolerance intervals (Guttman 1970):

- A *β-content* tolerance interval with **confidence level** $(1-\alpha)100\%$ is constructed so that it contains *at least* $\beta100\%$ of the population (i.e., the coverage is at least $\beta100\%$) with probability $(1-\alpha)100\%$.
- A *β-expectation* tolerance interval is constructed so that it contains *on average* $\beta100\%$ of the population (i.e., the average coverage is $\beta100\%$).

A β-expectation tolerance interval with coverage $\beta100\%$ is equivalent to a prediction interval for $k = 1$ future observation with associated confidence level $\beta100\%$. Note that there is no explicit confidence level associated with a β-expectation tolerance interval. If a β-expectation tolerance interval is treated as a β-content tolerance interval, the confidence level associated with this tolerance interval is usually around 50% (e.g., Guttman 1970). Thus, a β-content tolerance interval with coverage $\beta100\%$ will usually be wider than a β-expectation tolerance interval with the same coverage if the confidence level associated with the β-content tolerance interval is more than 50%.

It can be shown (e.g., Conover 1980) that an upper confidence interval for the pth quantile with confidence level $(1-\alpha)100\%$ is equivalent to an upper β-content tolerance interval with coverage $100p\%$ and confidence level $(1-\alpha)100\%$. Also, a lower confidence interval for the pth quantile with confidence level $(1-\alpha)100\%$ is equivalent to a lower β-content tolerance interval with coverage $100(1-p)\%$ and confidence level $(1-\alpha)100\%$.

Tolerance intervals have long been applied to quality control and life testing problems. In environmental monitoring, USEPA has in the past proposed using tolerance intervals in at least two different ways: compliance-to-background comparisons and compliance-to-fixed standard comparisons. However, current guidance (USEPA 2009) recommends using prediction intervals or confidence intervals in place of tolerance intervals except in the case when concentrations at compliance wells need to be compared to a groundwater protection standard (GWPS) and background concentrations are themselves above the GWPS (see USEPA 2009, pp. 6–46, 7–21).

6.4.1   *Tolerance Intervals for a Normal Distribution*

Section 5.3.1 contains an example in which an MCL of 30 ppb for aldicarb should not be exceeded more than 5% of the time. Using the data from three groundwater monitoring compliance wells (four monthly samples at each well) stored in `EPA.09.Ex.21.1.aldicarb.df`, first we assume the facility is in compliance/assessment monitoring, so we compute a lower 99% confidence limit for the 95th percentile for the distribution at each of the three compliance wells, yielding

25.3, 25.7, and 5.5 ppb.   This is equivalent to computing a lower β-content tolerance limit with coverage 5 % and associated confidence level of 99 %. Here are the results for the first well:

```
> Aldicarb <- EPA.09.Ex.21.1.aldicarb.df$Aldicarb.ppb

> Well <- EPA.09.Ex.21.1.aldicarb.df$Well

> tolIntNorm(Aldicarb[Well == "Well.1"], coverage = 0.05,
    ti.type = "lower", conf.level = 0.99)

Results of Distribution Parameter Estimation
--------------------------------------------


Assumed Distribution:            Normal

Estimated Parameter(s):          mean = 23.10000
                                 sd   =  4.93491

Estimation Method:               mvue

Data:                            Aldicarb[Well == "Well.1"]

Sample Size:                     4

Tolerance Interval Coverage:     5%

Coverage Type:                   content

Tolerance Interval Method:       Exact

Tolerance Interval Type:         lower

Confidence Level:                99%

Tolerance Interval:              LTL = 25.28550
                                 UTL =      Inf
```

Here are the results for wells 2 and 3:

```
> tolIntNorm(Aldicarb[Well = ="Well.2"], coverage = 0.05,
    ti.type = "lower",
    conf.level = 0.99)$interval$limits["LTL"]

    LTL
25.66086
```

```
> tolIntNorm(Aldicarb[Well == "Well.3"], coverage = 0.05,
    ti.type = "lower",
    conf.level = 0.99)$interval$limits["LTL"]

   LTL
5.45563
```

Instead of calling `tolIntNorm` three separate times, you can instead just use the following single command:

```
> sapply(split(Aldicarb, Well), function(x) {
    tolIntNorm(x, coverage = 0.05, ti.type = "lower",
    conf.level = 0.99)$interval$limits["LTL"]})

Well.1.LTL Well.2.LTL Well.3.LTL
  25.28550   25.66086    5.45563
```

Since none of the LTLs is above the 30 ppb MCL, no corrective action is needed. In the second part of the example, however, we assume the site is in corrective action monitoring, and we compute the one-sided 99 % *upper* confidence limit for the 95th percentile and compare that to the MCL. This is equivalent to computing an upper β-content tolerance limit with coverage 95 % and associated confidence level of 99 %.

```
> sapply(split(Aldicarb, Well), function(x) {
    tolIntNorm(x, coverage = 0.95, ti.type = "upper",
    conf.level = 0.99)$interval$limits["UTL"]})

Well.1.UTL Well.2.UTL Well.3.UTL
  67.92601   45.38336   23.61286
```

In this case there is evidence that corrective action is still needed at Wells 1 and 2 since the UTL is greater than 30 ppb. Of course, a major problem in this whole example is the very small sample size at each well ($n = 4$) used to compute the intra-well tolerance limit.

6.4.2   Tolerance Intervals for a Lognormal Distribution

In Sect. 6.2.2 we computed a prediction interval to compare the TcCB concentrations between the Cleanup and Reference areas (see Figs. 1.1, 1.2, and 1.3 in Sect. 1.11.3). Based on the data from the Background area, the one-sided upper 95 % prediction limit for the next $k = 77$ observations (there are 77 observations in the Cleanup area) is 2.68 ppb. There are seven observations in the Cleanup area larger than 2.68, so the prediction interval indicates residual contamination is present in the Cleanup area.

Some guidance documents suggest constructing a one-sided upper tolerance interval based on the Reference area and comparing all of the observations from the Cleanup area to the upper tolerance limit. This is sometimes called the "Hot-Measurement Comparison" (USEPA 1994b). Millard et al. (2014) explain why this method should never be used because you do not know the true Type I error rate. In this case, the one-sided upper 95 % β-content tolerance limit with associated confidence level 95 % is 1.42 ppb (versus 2.68 ppb for the upper prediction limit).

```
> with(EPA.94b.tccb.df,
    tolIntLnorm(TcCB[Area == "Reference"], coverage = 0.95,
    ti.type = "upper", conf.level = 0.95))

Results of Distribution Parameter Estimation
--------------------------------------------


Assumed Distribution:            Lognormal

Estimated Parameter(s):          meanlog = -0.6195712
                                 sdlog   =  0.4679530

Estimation Method:               mvue

Data:                            TcCB[Area == "Reference"]

Sample Size:                     47

Tolerance Interval Coverage:     95%

Coverage Type:                   content

Tolerance Interval Method:       Exact

Tolerance Interval Type:         upper

Confidence Level:                95%
Tolerance Interval:              LTL = 0.000000
                                 UTL = 1.424970
```

Example 17-3 of USEPA (2009, p. 17-17) presents a similar example using groundwater monitoring data on chrysene concentrations from two background wells and three compliance wells. In **EnvStats** theses data are stored in EPA.09.Ex.17.3.chrysene.df.

```
> EPA.09.Ex.17.3.chrysene.df

    Month    Well  Well.type Chrysene.ppb
1        1 Well.1 Background         19.7
2        2 Well.1 Background         39.2
3        3 Well.1 Background          7.8
...
```

18	2 Well.5 Compliance	30.5
19	3 Well.5 Compliance	15.0
20	4 Well.5 Compliance	23.4

A check on the distribution of the background well concentrations indicates the data are right-skewed and can be modeled with a lognormal distribution. The one-sided upper 95 % β-content tolerance limit with associated confidence level 95 % is 90.9 ppb, and since none of the concentrations at the compliance wells is larger than this there is no evidence of contamination. Again, this method should never be used because you do not know the true Type I error rate.

```
> with(EPA.09.Ex.17.3.chrysene.df,
    tolIntLnorm(Chrysene.ppb[Well.type == "Background"],
    coverage = 0.95, ti.type = "upper",
    conf.level = 0.95))$interval$limits["UTL"]

    UTL
90.9247
```

6.4.3   Tolerance Intervals for a Gamma Distribution

You can use the EnvStats functions `tolIntGamma` or `tolIntGammaAlt` to construct tolerance intervals assuming a gamma distribution. The upper 95 % tolerance interval based on the Reference area TcCB data is 1.33 ppb, as opposed to is 1.42 ppb assuming a lognormal distribution:

```
> with(EPA.94b.tccb.df,
    tolIntGamma(TcCB[Area == "Reference"],
    coverage = 0.95, ti.type = "upper",
    conf.level = 0.95))$interval$limits["UTL"]

    UTL
1.325023
```

6.4.4   Nonparametric Tolerance Intervals

You can construct tolerance intervals without making any assumption about the distribution of the background data, except that the distribution is continuous. These kinds of tolerance intervals are called *nonparametric tolerance intervals*. Of course, nonparametric tolerance intervals still require the assumption that the distribution of future observations is the same as the distribution of the observations used to create the tolerance interval. Just as for nonparametric prediction intervals, nonparametric tolerance intervals are based on the ranked data.

Example 17-4 of USEPA (2009, pp. 17–21) contains copper concentration data from three background wells and two compliance wells as shown in Table 6.7 below:

		Background		Compliance	
Month	**Well 1**	**Well 2**	**Well 3**	**Well 4**	**Well 5**
1	<5	9.2	<5		
2	<5	<5	5.4		
3	7.5	<5	6.7		
4	<5	6.1	<5		
5	<5	8.0	<5	6.2	<5
6	<5	5.9	<5	<5	<5
7	6.4	<5	<5	7.8	5.6
8	6.0	<5	<5	10.4	<5

Table 6.7 Copper concentrations (ppb) from groundwater monitoring wells

In EnvStats these data are stored in `EPA.09.Ex.17.4.copper.df`:

```
> EPA.09.Ex.17.4.copper.df

   Month   Well  Well.type Copper.ppb.orig Copper.ppb Censored
1      1 Well.1 Background             <5        5.0     TRUE
2      2 Well.1 Background             <5        5.0     TRUE
3      3 Well.1 Background            7.5        7.5    FALSE
...
38     6 Well.5 Compliance            <5        5.0     TRUE
39     7 Well.5 Compliance           5.6        5.6    FALSE
40     8 Well.5 Compliance            <5        5.0     TRUE
```

Because of the large percentage of non-detects, a nonparametric approach is used. In this example the 95 % confidence upper tolerance limit is computed using the maximum value of the background wells (i.e., 9.2 ppb) with the idea that this limit will be used as a threshold value for concentrations observed in the two compliance wells (i.e., if any concentrations at the compliance wells exceed this limit, this indicates there may be contamination in the groundwater). This is the "Hot-Measurement Comparison," and as already discussed above there are some major problems with this technique.

```
> with(EPA.09.Ex.17.4.copper.df,
    tolIntNpar(Copper.ppb[Well.type == "Background"],
    conf.level = 0.95, ti.type = "upper", lb = 0)

Results of Distribution Parameter Estimation
--------------------------------------------

Assumed Distribution:            None

Data:
        Copper.ppb[Well.type == "Background"]

Sample Size:                     24
```

```
Tolerance Interval Coverage:        88.26538%

Coverage Type:                      content

Tolerance Interval Method:          Exact

Tolerance Interval Type:            upper

Confidence Level:                   95%

Tolerance Limit Rank(s):            24

Tolerance Interval:                 LTL = 0.0
                                    UTL = 9.2
```

There is evidence of possible contamination at Well 4 because of the one value greater than 9.2 ppb. However, the coverage associated with a nonparametric tolerance interval based on $n = 24$ observations is only 88 %, so the probability of a Type I error using this method is 12 %. In order to obtain coverage of at least 95 %, you would need 59 background samples:

```
> tolIntNparN(ti.type = "upper", coverage = 0.95,
    conf.level = 0.95)

[1] 59
```

6.5 Summary

- Any activity that requires comparing new values to "background" or "standard" values creates a decision problem: if the new values greatly exceed the background or standard value, is this evidence of a true difference (i.e., is there contamination)?
- Statistical tests are used as objective tools to decide whether a change has occurred. For a monitoring program that involves numerous tests over time, figuring out how to balance the **overall** Type I error with the power of detecting a change is a multiple comparisons problem.
- Prediction intervals and tolerance intervals are two tools that you can use to attempt to solve the multiple comparisons problem.
- Table 6.1 lists the functions available in ENVSTATS for constructing prediction intervals.
- Table 6.6 lists the functions available in ENVSTATS for constructing tolerance intervals.

Chapter 7

Hypothesis Tests

7.1 Introduction

If you are comparing chemical concentrations between a background area and a potentially contaminated area, how different do the concentrations in these two areas have to be before you decide that the potentially contaminated area is in fact contaminated? In the last chapter we showed how to use prediction and tolerance intervals to try to answer this question. There are other kinds of hypothesis tests you can use as well. R contains several functions for performing classical statistical hypothesis tests, such as t-tests, analysis of variance, linear regression, nonparametric tests, quality control procedures, and time series analysis (see the R documentation and help files). ENVSTATS contains modifications of some of these functions (e.g., `summaryStats` and `stripChart`), as well as functions for statistical tests that are not included in R but that are used in environmental statistics, such as the Shapiro-Francia goodness-of-fit test, Kendall's seasonal test for trend, and the quantile test for a shift in the tail of the distribution (see the help file *Hypothesis Tests*). This chapter discusses these functions. See Millard et al. (2014) for a more in-depth discussion of hypothesis tests.

7.2 Goodness-of-Fit Tests

Most commonly used parametric statistical tests assume the observations in the random sample come from a normal population. In fact, the usual assumptions are that the observations are independent, the variance of the distribution is constant, and the distribution is normal. These three assumptions are listed in decreasing importance with respect to maintaining the assumed Type I error (van Belle 2008; Millard et al. 2014). So how do you know whether the assumption of a normal distribution is valid? We saw in Chap. 3 how to make a visual assessment of this assumption using Q-Q plots. Another way to verify this assumption is with a goodness-of-fit test, which lets you specify what kind of distribution you think the data come from and then compute a test statistic and a p-value.

A goodness-of-fit test may be used to test the null hypothesis that the data come from a specific distribution, such as "the data come from a normal distribution with mean 10 and standard deviation 2," or to test the more general null hypothesis that the data come from a particular family of distributions, such as "the data come from a lognormal distribution." Goodness-of-fit tests are mostly used to test the latter kind of hypothesis, since in practice we rarely know or want to specify the parameters of the distribution.

In practice, goodness-of-fit tests may be of limited use for very large or very small sample sizes. Almost any goodness-of-fit test will reject the null hypothesis of the specified distribution if the number of observations is very large, since "real" data are never distributed according to any theoretical distribution (Conover 1980). On the other hand, with only a very small number of observations, no test will be able to determine whether the observations appear to come from the hypothesized distribution or some other totally different looking distribution.

Function	Description
gofTest	Shapiro-Wilk, Shapiro-Francia, probability plot correlation coefficient (PPCC), and zero-skew goodness-of-fit tests for a normal, lognormal, three-parameter lognormal, zero-modified normal, or zero-modified lognormal (delta) distribution
	Wilk-Shapiro test for a uniform [0, 1] distribution
	Shapiro-Wilk type test for *any* continuous distribution available in EnvStats
	PPCC goodness-of-fit test for extreme value distribution
	Kolmogorov-Smirnov goodness-of-fit test to compare a sample with a specified probability distribution or to compare two samples
	Chi-square goodness-of-fit test for a specified probability distribution
gofGroupTest	Shapiro-Wilk, Shapiro-Francia, and PPCC goodness-of-fit tests for normality for two or more groups
gofCensoredTest	Shapiro-Wilk, Shapiro-Francia, and PPCC goodness-of-fit tests for normality for censored data

Table 7.1 Functions in ENVSTATS for goodness-of-fit tests

Table 7.1 shows the functions available in **EnvStats** for performing goodness-of-fit tests. The function gofTest lets you perform the one-sample Shapiro-Wilk, Shapiro-Francia, or probability plot correlation coefficient (PPCC) goodness-of-fit test for normality. You can also use this function to determine whether a set of observations appears to come from a lognormal, three-parameter lognormal, zero-modified normal, or zero-modified lognormal (delta) distribution. In addition, you can perform a Shapiro-Wilk type test to test for *any* continuous distribution that is available in EnvStats (see the help file for Distribution.df). This function also lets you perform the one-sample PPCC test for the extreme value distribution, as well as the Kolmogorov-Smirnov and chi-square goodness-of-fit tests to compare a sample with any specified probability distribution. The function gofGroupTest lets you test the assumption of normality for several groups of data simultaneously while controlling the overall Type I error.

The function `gofCensoredTest` lets you test the assumption of normality based on censored data, and the discussion of this function is deferred until the next chapter. There are specific printing and plotting methods associated with the results of `gofTest`, `gofGroupTest`, and `gofCensoredTest`.

7.2.1   One-Sample Goodness-of-Fit Tests for Normality

In Chaps. 1 and 3 we saw that the Reference area TcCB data appear to come from a lognormal distribution based on a histogram (Fig. 1.2), an empirical cdf plot (Fig. 1.5), a normal Q-Q plot (Fig. 1.7), a Tukey mean-difference Q-Q plot (Fig. 1.8), and a plot of the probability plot correlation coefficient (PPCC) versus λ for a variety of Box-Cox transformations (Fig. 3.7). In Sect. 1.11.7 we showed the results of using the Shapiro-Wilk test to test the adequacy of the lognormal distribution. Here we will formally test whether the Reference area TcCB data appear to come from a normal distribution versus a lognormal distribution, and in the call to `gofTest` for testing lognormality we will specify using the alternative parameterization of the lognormal distribution (i.e., estimating the mean and CV of the original distribution).

```
> attach(EPA.94b.tccb.df)

> TcCB.Ref <- TcCB[Area == "Reference"]

> sw.list.norm <- gofTest(TcCB.Ref)

> sw.list.norm

Results of Goodness-of-Fit Test
-------------------------------

Test Method:                   Shapiro-Wilk GOF

Hypothesized Distribution:     Normal

Estimated Parameter(s):        mean = 0.5985106
                               sd   = 0.2836408

Estimation Method:             mvue

Data:                          TcCB.Ref

Sample Size:                   47

Test Statistic:                W = 0.9176408

Test Statistic Parameter:      n = 47
```

P-value: 0.002768207

Alternative Hypothesis: True cdf does not equal the
                                        Normal Distribution.

```
> sw.list.lnormAlt <- gofTest(TcCB.Ref, dist = "lnormAlt")
```

```
> sw.list.lnormAlt
```

Results of Goodness-of-Fit Test

Test Method: Shapiro-Wilk GOF

Hypothesized Distribution: Lognormal

Estimated Parameter(s):   mean = 0.5989072
                                        cv   = 0.4899539

Estimation Method: mvue

Data: TcCB.Ref

Sample Size: 47

Test Statistic: W = 0.978638

Test Statistic Parameter: n = 47

P-value: 0.5371935

Alternative Hypothesis: True cdf does not equal the
                                        Lognormal Distribution.

The p-value for the test of normality ($p = 0.003$) clearly indicates that we should not assume the Reference area TcCB data come from a normal distribution, but the assumption of a lognormal distribution appears to be adequate ($p = 0.54$). Figures 7.1 and 7.2 show companion plots for the results of the Shapiro-Wilk tests for normality and lognormality, respectively. These plots include the observed distribution overlaid with the fitted distribution, the observed and fitted CDF, the normal Q-Q plot, and the results of the hypothesis test. They were created with these commands:

```
> plot(sw.list.norm)
```

```
> plot(sw.list.lnormAlt)
```

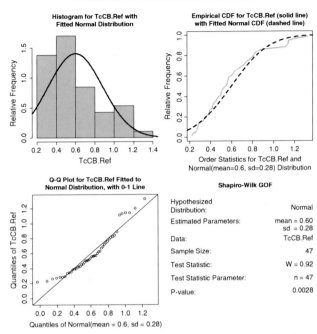

Fig. 7.1 Companion plots for the Shapiro-Wilk test for normality for the Reference area TcCB data

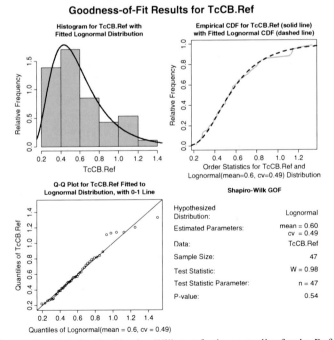

Fig. 7.2 Companion plots for the Shapiro-Wilk test for lognormality for the Reference area TcCB data

You can use the `plot.type` argument to specify particular plots. For example, to generate a Tukey Mean-Difference Q-Q plot for the test of normality, you would type:

```
> plot(sw.list.norm, plot.type = "Tukey")
```

7.2.2   Testing Several Groups for Normality

If you have several sets of observations you want to test for normality, you may encounter the multiple comparisons problem. For example, regulations for monitoring groundwater at hazardous and solid waste sites may require performing statistical analyses even when there are only small sample sizes at each monitoring well. As we noted above, goodness-of-fit tests are not very useful with small sample sizes; there is simply not enough information to determine whether the data appear to come from the hypothesized distribution or not. Gibbons (1994) suggests pooling the measures from several upgradient wells to establish "background." Due to spatial variability, the wells may have different means and variances, yet you would like to test the assumption of a normal distribution for the chemical concentration at each of the upgradient wells.

Wilk and Shapiro (1968) suggest two different test statistics for the problem of testing the normality of K separate groups, using the results of the Shapiro-Wilk test applied to random samples from each of the K groups. Both test statistics are functions of the K p-values that result from performing the test on each of the K samples.   Under the null hypothesis that all K samples come from normal distributions, the p-values represent a random sample from a uniform distribution on the interval [0,1].   Since these two test statistics are based solely on the p-values, they are really meta-analysis statistics (Fisher and van Belle 1993), and can be applied to the problem of combining the results from K independent hypothesis tests, where the hypothesis tests are not necessarily goodness-of-fit tests. You can use the function `gofGroupTest` for performing group tests for normality.

Example 10-4 of USEPA (2009, pp. 10–20) involves looking at observations of nickel concentrations (ppb) collected over 5 months at 4 monitoring wells. These data are stored in the data frame `EPA.09.Ex.10.1.nickel.df`.

```
> EPA.09.Ex.10.1.nickel.df
```

	Month	Well	Nickel.ppb
1	1	Well.1	58.8
2	3	Well.1	1.0
...			
19	8	Well.4	10.0
20	10	Well.4	637.0

Figure 7.3 displays the observations for each well, and Fig. 7.4 displays the log-transformed observations, created with these commands:

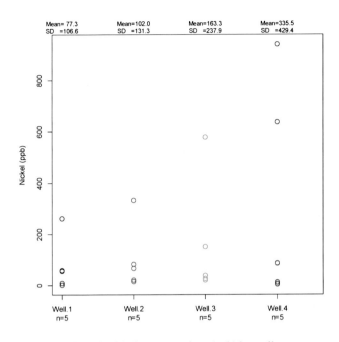

Fig. 7.3 Nickel concentrations (ppb) by well

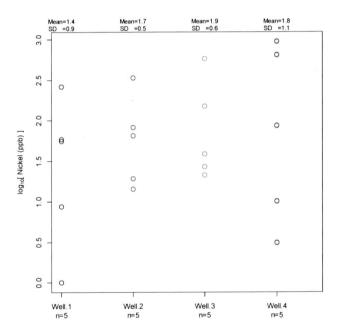

Fig. 7.4 Log-transformed nickel concentrations by well

```
> stripChart(Nickel.ppb ~ Well, col = 1:4,
    data = EPA.09.Ex.10.1.nickel.df, show.ci = FALSE,
    ylab = "Nickel (ppb)")

> stripChart(log10(Nickel.ppb) ~ Well, col = 1:4, data =
    EPA.09.Ex.10.1.nickel.df, show.ci = FALSE,
    ylab = expression(paste(log[10], "[ Nickel (ppb) ]")))
```

First we will use the Shapiro-Wilk group test to test the null hypotheses that the observations at each well represent a sample from some kind of normal distribution, but the population means and/or variances may differ between wells.

```
> sw.list <- gofGroupTest(Nickel.ppb ~ Well,
    data = EPA.09.Ex.10.1.nickel.df)

> sw.list
```

```
Results of Group Goodness-of-Fit Test
-------------------------------------

Test Method:                     Wilk-Shapiro GOF (Normal Scores)

Hypothesized Distribution:       Normal

Data:                            Nickel.ppb

Grouping Variable:               Well

Data Source:                     EPA.09.Ex.10.1.nickel.df

Number of Groups:                4

Sample Sizes:                    Well.1 = 5
                                 Well.2 = 5
                                 Well.3 = 5
                                 Well.4 = 5

Test Statistic:                  z (G) = -3.658696

P-values for
Individual Tests:                Well.1 = 0.03510747
                                 Well.2 = 0.02385344
                                 Well.3 = 0.01120775
                                 Well.4 = 0.10681461

P-value for
Group Test:                      0.0001267509
```

| Alternative Hypothesis: | At least one group does not come from a Normal Distribution. |

Now we'll do the same test but assume a lognormal distribution:

```
> sw.log.list <- gofGroupTest(Nickel.ppb ~ Well,
    data = EPA.09.Ex.10.1.nickel.df, dist = "lnorm")

> sw.log.list
```

Results of Group Goodness-of-Fit Test

Test Method:	Wilk-Shapiro GOF (Normal Scores)
Hypothesized Distribution:	Lognormal
Data:	Nickel.ppb
Grouping Variable:	Well
Data Source:	EPA.09.Ex.10.1.nickel.df
Number of Groups:	4
Sample Sizes:	Well.1 = 5
	Well.2 = 5
	Well.3 = 5
	Well.4 = 5
Test Statistic:	z (G) = 0.240172
P-values for Individual Tests:	Well.1 = 0.6898164
	Well.2 = 0.6700394
	Well.3 = 0.3208299
	Well.4 = 0.5041375
P-value for Group Test:	0.5949015
Alternative Hypothesis:	At least one group does not come from a Lognormal Distribution.

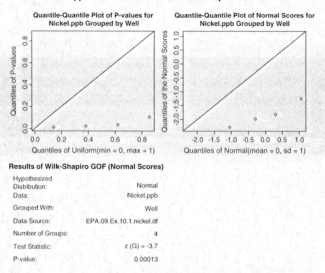

Fig. 7.5 Companion plots for Shapiro-Wilk group goodness-of-fit test for normality of the nickel concentrations

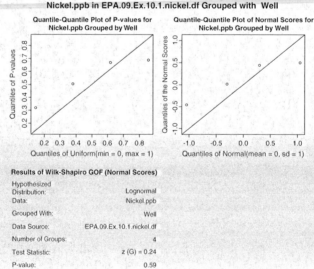

Fig. 7.6 Companion plots for Shapiro-Wilk group goodness-of-fit test for lognormality of the nickel concentrations

Figures 7.5 and 7.6 display the companion plots for each of the hypotheses we tested. They were created with these commands:

```
> plot(sw.list)
```

```
> plot(sw.log.list)
```

There is clear evidence that the raw concentrations do not come from normal distributions, but the assumption of lognormal distributions appears to be adequate.

7.2.3   One-Sample Goodness-of-Fit Tests for Other Distributions

Section 1.11.7 in Chap. 1 showed how to test for the assumption of a gamma distribution for the Reference Area TcCB data. In general, you can perform a goodness-of-fit test for any continuous distribution available in EnvStats by using the dist argument to gofTest.

Three other commonly used goodness-of-fit tests are the Kolmogorov-Smirnov goodness-of-fit test (Zar 2010), the chi-square goodness-of-fit test (Zar 2010), and the probability plot correlation coefficient (PPCC) goodness-of-fit test (Filliben 1975; Vogel 1986). The function gofTest uses the Kolmogorov-Smirnov and chi-square tests built into R (i.e., ks.test and chisq.test). ENVSTATS also adds the PPCC test for the extreme value distribution. The help file for gofTest contains more detailed information and examples.

In Chap. 3, Sect. 3.7.2, we created a Q-Q plot to determine whether a set of benzene concentrations appear to come from a Poisson distribution. Out of the 36 observations, 33 are reported as "<2", and these observations were set to half the detection limit (i.e., 1). We can use the chi-square goodness-of-fit test to formally test whether these data appear to come from a Poisson distribution. When using the chi-square test to test whether data appear to come from a discrete distribution, you have to supply the vector of cut points that define the bins you want to use. Each bin is defined as all values greater than the lower cut point and less than or equal to the upper cut point. Here we will use the cut points -1, 0, 2, and ∞, which correspond to bins that hold 14 %, 55 %, and 31 % of the distribution based on the estimated parameter $\lambda = 1.94$.

```
> attach(EPA.92c.benzene1.df)
```

```
> Benzene[Censored] <- 1
```

```
> table(Benzene)
```

```
Benzene
 1 10 12 15
33  1  1  1
```

```
> lambda.hat <- epois(Benzene)$parameters
```

```
> lambda.hat

  lambda
1.944444

> ppois(c(-1, 0, 2, Inf), lambda = lambda.hat)

[1] 0.0000000 0.1430667 0.6917097 1.0000000

> diff(ppois(c(-1, 0, 2, Inf), lambda = lambda.hat))

[1] 0.1430667 0.5486431 0.3082903

> chisq.list <- gofTest(Benzene, test = "chisq",
    dist = "pois", cut.points = c(-1, 0, 2, Inf))

> chisq.list
```

```
Results of Goodness-of-Fit Test
-------------------------------


Test Method:                    Chi-square GOF

Hypothesized Distribution:      Poisson

Estimated Parameter(s):         lambda = 1.944444

Estimation Method:              mle/mme/mvue

Data:                           Benzene

Sample Size:                    36

Test Statistic:                 Chi-square = 19.94695

Test Statistic Parameter:       df = 1

P-value:                        7.962074e-06

Alternative Hypothesis:         True cdf does not equal the
                                Poisson Distribution.
```

The p-value is essentially 0, so we have evidence that the assumption of a Poisson distribution is not valid. Figure 7.7 shows companion plots to the test, created with this command:

```
> plot(chisq.list)
```

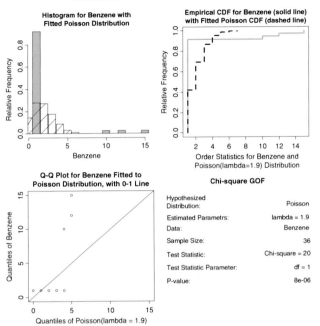

Fig. 7.7 Companion plots for chi-square goodness-of-fit test of Poisson distribution for benzene concentrations

7.2.4   Two-Sample Goodness-of-Fit Test to Compare Samples

You can use the Kolmogorov-Smirnov test to test whether two sets of observations appear to come from the exact same distribution. In Chap. 3, Sect. 3.7.3, we created a Q-Q plot comparing the Reference area and Cleanup area TcCB concentrations based on the log-transformed data. The Kolmogorov-Smirnov test yields a p-value of 0.013, so there is evidence that these two distributions differ (Fig. 7.8).

```
> attach(EPA.94b.tccb.df)

> log.TcCB.ref <- log(TcCB[Area=="Reference"])

> log.TcCB.clean <- log(TcCB[Area=="Cleanup"])

> ks.list <- gofTest(x = log.TcCB.ref, y = log.TcCB.clean,
    test = "ks")

> ks.list

Results of Goodness-of-Fit Test
-------------------------------

Test Method:                  2-Sample K-S GOF
```

```
Hypothesized Distribution:      Equal

Data:                           log.TcCB.ref and log.TcCB.clean

Number NA/NaN/Inf's Removed:    0 and 0

Sample Sizes:                   n.x = 47
                                n.y = 77

Test Statistic:                 ks = 0.2821221
Test Statistic Parameters:      n = 47
                                m = 77

P-value:                        0.01920147

Alternative Hypothesis:         The cdf of 'x' does not equal
                                the cdf of 'y'.

> plot(ks.list)
```

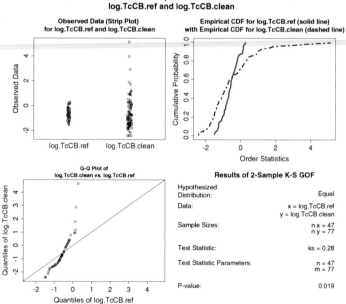

Fig. 7.8 Companion plots for Kolmogorov-Smirnov goodness-of-fit test comparing log-transformed TcCB data for reference and cleanup areas

7.3   One-, Two-, and *k*-Sample Comparison Tests

Frequently in environmental studies, we are interested in comparing concentrations to a standard, or comparing concentrations between two or more areas (e.g., background versus potentially contaminated). R comes with several built-in functions for performing standard hypothesis tests for one-, two-, and *k*-sample comparisons (e.g., Student's t-test, analysis of variance, etc.). Table 7.2 lists additional functions available in ENVSTATS for comparing samples. See the help files and Millard et al. (2014) for more detailed discussion of these functions. In this section we'll give examples of using the summaryStats and stripChart functions to perform 2- and *k*-sample comparisons, as well as an example of performing Chen's modified t-test, and comparing a linear rank test with the quantile test.

Comparison Type	Function	Description
Location	summaryStats	Summary statistics, p-values, and confidence intervals for mean or pseudo-median
	stripChart	Strip chart with confidence intervals for mean or pseudo-median
	chenTTest	Chen's modified one-sample or paired t-test for skewed data
	oneSamplePermutationTest	Fisher's one-sample permutation test for location
	twoSamplePermutationTestLocation	Two-sample or paired permutation test to compare locations
	signTest	Sign test for one-sample or paired data
	twoSampleLinearRankTest	Two-sample linear rank test
Quantile	quantileTest	Quantile test for a shift in the tail
Proportion	twoSamplePermutationTestProportion	Two-sample or paired permutation test to compare proportions
Variance	varTest	One-sample chi-square test on variance, or two-sample F test for equal variances
	varGroupTest	Levene's or Bartlett's test for equal variances among *k* populations

Table 7.2 Functions in ENVSTATS for comparison tests

7.3.1 Two- and k-Sample Comparisons for Location

The EnvStats function summaryStats not only displays basic summary statistics in a nice format, but it also allows you to perform standard tests for two-sample or *k*-sample comparisons. In Chap. 1 we looked at summary statistics and plots comparing the TcCB data in the Reference and Cleanup areas. Here we'll perform some hypothesis tests to compare the concentrations in these two areas. Here are the standard two-sample t-test results for the log-transformed concentrations:

```
> summaryStats(log(TcCB) ~ Area, data = EPA.94b.tccb.df,
      digits = 2, p.value = TRUE, stats.in.rows = TRUE)
```

	Cleanup	Reference	Combined
N	77	47	124
Mean	-0.55	-0.62	-0.57
SD	1.36	0.47	1.11
Median	-0.84	-0.62	-0.72
Min	-2.41	-1.51	-2.41
Max	5.13	0.29	5.13
p.value.between			0.73
95%.LCL.between			-0.48
95%.UCL.between			0.34

In Chap. 1 we noted that most of the observations in the Cleanup area are comparable to (or even smaller than) the observations in the Reference area, but there are a few very large "outliers" in the Cleanup area. By default, when the argument p.value is set to TRUE and there are two groups, the p-value is based on Student's t-test assuming equal variances (which is *not* the default behavior of the built in R function t.test). In this example the standard deviation in the Cleanup area is more than twice the standard deviation in the Reference area, so now we'll perform the test allowing for different variances in the two groups:

```
> summaryStats(log(TcCB) ~ Area, data = EPA.94b.tccb.df,
      digits = 2, p.value = TRUE, stats.in.rows = TRUE,
      test.arg.list = list(var.equal = FALSE))
```

	Cleanup	Reference	Combined
N	77	47	124
Mean	-0.55	-0.62	-0.57
SD	1.36	0.47	1.11
Median	-0.84	-0.62	-0.72
Min	-2.41	-1.51	-2.41
Max	5.13	0.29	5.13
Welch.p.value.between			0.67
95%.LCL.between			-0.41
95%.UCL.between			0.26

Even allowing for different variances, the t-test does not provide evidence of a difference in the mean value between the two areas. When there are more than two groups, by default a standard analysis of variance F-test is performed. Instead of performing the t-test or F-test, you can set `test="nonparametric"` to perform the Wilcoxon or Kruskal-Wallis rank sum test.

In Chap. 1 we used the function `stripChart` to produce one-dimensional scatterplots of the log-transformed TcCB data by area, along with confidence intervals for the means (Fig. 1.1). You can also display the results of testing for a difference between the two means (Fig. 7.9):

```
> stripChart(log(TcCB) ~ Area, data = EPA.94b.tccb.df,
    col = c("red", "blue"), p.value = TRUE,
    ylab = "Log [ TcCB (ppb) ]")
```

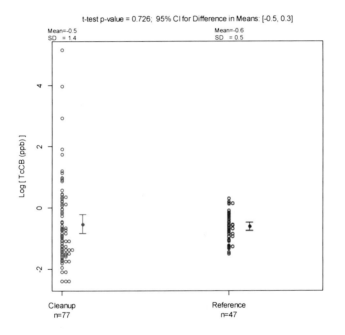

Fig. 7.9 One-dimensional scatterplots, 95 % confidence intervals for the means, and results of Student's t-test comparing TcCB concentrations at Reference and Cleanup areas

As with `summaryStats`, when you set `p.value=TRUE` in the call to `stripChart` and there are two groups, the standard two-sample t-test is performed, and when there are more than two groups a standard analysis of variance F-test is performed. Setting `test="nonparametric"` will compute the Wilcoxon or Kruskal-Wallis rank sum test instead.

7.3.2   Chen's Modified One-Sample t-Test for Skewed Data

Student's t-test, Fisher's one-sample permutation test, and the Wilcoxon signed rank test all assume that the underlying distribution is symmetric about its mean. Chen (1995b) developed a modified t-statistic for performing a one-sided test of hypothesis on the mean of a skewed distribution. For the case of a positively skewed distribution, her test can be applied to the upper one-sided alternative hypothesis (i.e., H_0: $\mu \le \mu_0$ vs. H_a: $\mu > \mu_0$). For the case of a negatively skewed distribution, her test can be applied to the lower one-sided alternative hypothesis (i.e., H_0: $\mu \ge \mu_0$ vs. H_a: $\mu < \mu_0$). Since environmental data are usually positively skewed, her test would usually be applied to the case of testing the upper one-sided hypothesis.

The guidance document *Calculating Upper Confidence Limits for Exposure Point Concentrations at Hazardous Waste Sites* (USEPA 2002d, Exhibit 9, p. 16) contains an example of 60 observations from an exposure unit (Fig. 7.10). In ENVSTATS these data are stored in the vector EPA.02d.Ex.9.mg.per.L.vec.

```
> hist(EPA.02d.Ex.9.mg.per.L.vec, col = "cyan",
    xlab = "Concentration (mg/L)", main = "")
```

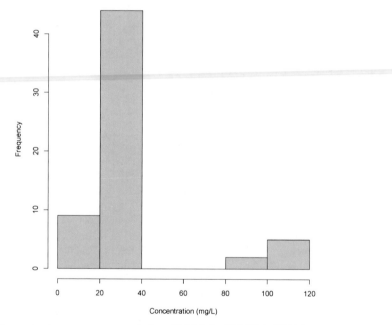

Fig. 7.10 Concentrations at exposure unit from Exhibit 9 of USEPA (2002d)

The Shapiro-Wilk goodness-of-fit test rejects the null hypothesis of a normal distribution and a lognormal distribution.

```
> gofTest(EPA.02d.Ex.9.mg.per.L.vec)$p.value
```

```
[1] 2.496781e-12
```

```
> gofTest(EPA.02d.Ex.9.mg.per.L.vec, dist = "lnorm")$p.value
```

```
[1] 3.349035e-09
```

In this example we will use Chen's modified t-test to test the null hypothesis that the average concentration is less than 30 mg/L versus the alternative that it is greater than 30 mg/L.

```
> chenTTest(EPA.02d.Ex.9.mg.per.L.vec, mu = 30)
```

```
Results of Hypothesis Test
--------------------------
```

```
Null Hypothesis:              mean = 30

Alternative Hypothesis:       True mean is greater than 30

Test Name:                    One-sample t-Test
                              Modified for
                              Positively-Skewed Distributions
                              (Chen, 1995)

Estimated Parameter(s):       mean = 34.566667
                              sd   = 27.330598
                              skew =  2.365778

Data:                         EPA.02d.Ex.9.mg.per.L.vec

Test Statistic:               t = 1.574075

Test Statistic Parameter:     df = 59

P-values:                     z               = 0.05773508
                              t               = 0.06040889
                              Avg. of z and t = 0.05907199

Confidence Interval for:      mean

Confidence Interval Method:   Based on z

Confidence Interval Type:     Lower

Confidence Level:             95%
```

```
Confidence Interval:              LCL =  29.82
                                  UCL =    Inf
```

The estimated mean, standard deviation, and skew are 35, 27, and 2.4, respectively. The p-value is 0.06, and the lower 95 % confidence interval is [29.8, ∞). Depending on what you use for your Type I error rate, you may or may not want to reject the null hypothesis.

The presentation of Chen's (1995b) method in USEPA (2002d) and Singh et al. (2010b, p. 52) is incorrect for two reasons: it is based on an intermediate formula instead of the actual statistic that Chen proposes, and it uses the intermediate formula to computer an *upper* confidence limit for the mean when the sample data are positively skewed. As explained above, for the case of positively skewed data, Chen's method is appropriate to test the upper one-sided alternative hypothesis that the population mean is greater than some specified value, and a one-sided upper alternative corresponds to creating a one-sided *lower* confidence limit, not an upper confidence limit (see, for example, Millard and Neerchal 2001, p. 371).

7.3.3 Two-Sample Linear Rank Tests and the Quantile Test

The Wilcoxon rank sum test is an example of a two-sample linear rank test. A linear rank test can be written as follows:

$$L = \sum_{i=1}^{n_1} a\left(R_{1i}\right) \tag{7.1}$$

where n_1 denotes the number of observations in group 1, R_{1i} denotes the rank of the *i*th observation in group 1, and $a()$ is some function that is called the **score function**. A linear rank test is based on the sum of the scores for group 1. For the Wilcoxon rank sum test, the function $a()$ is simply the identity function. Other functions may work better at detecting a small shift in location, depending on the shape of the underlying distributions. See the ENVSTATS help file for twoSampleLinearRankTest for more information.

The Wilcoxon rank sum test and other linear rank tests for shifts in location are all designed to detect a shift in the whole distribution of group 1 relative to the distribution of group 2. Sometimes, we may be interested in detecting a difference between the two distributions where only a portion of the distribution of group 1 is shifted relative to the distribution of group 2. The mathematical notation for this kind of shift is:

$$F_1(t) = (1-\varepsilon)\, F_2(t) + \varepsilon F_3(t) , \quad -\infty < t < \infty \tag{7.2}$$

where F_1 denotes the cumulative distribution function (cdf) of group 1, F_2 denotes the cdf of group 2, and ε denotes a fraction between 0 and 1. In the statistical literature, the distribution of group 1 is sometimes called a "contaminated" distribution, because it is the same as the distribution of group 2, except it is partially contaminated with another distribution. If the distribution of group 1 is partially shifted to the right of the distribution of group 2, F_3 denotes a cdf such that

$$F_3(t) \leq F_2(t) \;, \quad -\infty < t < \infty \tag{7.3}$$

with a strict inequality for at least one value of t. If the distribution of group 1 is partially shifted to the left of the distribution of group 2, F_3 denotes a cdf such that

$$F_3(t) \geq F_2(t) \;, \quad -\infty < t < \infty \tag{7.4}$$

with a strict inequality for at least one value of t.

The quantile test is a two-sample rank test to detect a shift in a proportion of one population relative to another population (Johnson et al. 1987). Under the null hypothesis, the two distributions are the same. If the alternative hypothesis is that the distribution of group 1 is partially shifted to the right of the distribution of group 2, the test combines the observations, ranks them, and computes k, which is the number of observations from group 1 out of the r largest observations. The test rejects the null hypothesis if k is too large.

In Chap. 1, Sect. 1.11.9, we compared the Reference area and Cleanup area TcCB concentrations using both the Wilcoxon rank sum test and the quantile test. The Wilcoxon rank sum test yields a p-value of 0.88, whereas the quantile test yields a p-value of 0.01. These results are not surprising, considering the histograms of the data shown in Fig. 1.2 in Sect. 1.11.3.

7.4   Testing for Serial Correlation

You can test for the presence of serial correlation in a time series or set of residuals from a linear fit using the ENVSTATS function `serialCorrelationTest`. You can test for the presence of lag-one serial correlation using either the rank von Neumann ratio test, the normal approximation based on the Yule-Walker estimate of lag-one correlation, or the normal approximation based on the MLE of lag-one correlation. Only the last method, however, allows for missing values in the time series. See the help file for `serialCorrelationTest` and Millard et al. (2014) for examples of testing for serial correlation.

7.5   Testing for Trend

Often in environmental studies we are interested in assessing the presence or absence of a long term trend. A parametric test for trend involves fitting a linear model that includes some measure of time as one of the predictor variables, and possibly allowing for serially correlated errors in the model. The ***Mann-Kendall test for trend*** (Mann 1945) is a nonparametric test for trend that does not assume normally distributed errors. Hirsch et al. (1982) introduced a modification of this test they call the ***seasonal Kendall test***. This test allows for seasonality and possibly serially correlated observations as well. Parametric and nonparametric tests for trend are described in detail in Millard et al. (2014). Table 7.3 lists the functions available in ENVSTATS for nonparametric tests for trend.

Function	Description
kendallTrendTest	Nonparametric test for monotonic trend based on Kendall's tau statistic
kendallSeasonalTrendTest	Nonparametric test for monotonic trend within each season based on Kendall's tau statistic. Allows for serial correlation as well

Table 7.3 Functions in ENVSTATS for nonparametric tests for trend

7.5.1   Testing for Trend in the Presence of Seasons

Figure 7.11 displays monthly estimated total phosphorus mass (mg) within a water column at station CB3.3e for the 5-year time period October 1984–September 1989 from a study on phosphorus concentration conducted in the Chesapeake Bay (Neerchal and Brunenmeister 1993). In ENVSTATS these data are stored in the data frame Total.P.df.

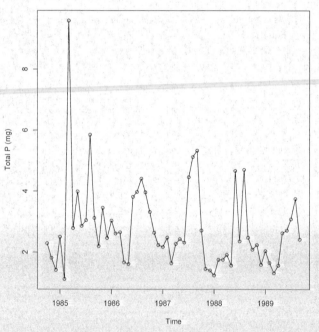

Fig. 7.11 Monthly estimated total phosphorus mass (mg) within a water column at station CB3.3e in the Chesapeake Bay

```
> Total.P.df

     CB3.1   CB3.3e Month Year
1    5.6330 2.30034   Oct 1984
2    3.6457 1.81900   Nov 1984
...
59   6.8030 3.75360   Aug 1989
60   6.2471 2.41853   Sep 1989
```

Figure 7.11 was created with these commands:

```
> with(Total.P.df, plot(CB3.3e, type = "o", xaxt = "n",
    xlab = "Time", ylab = "Total P (mg)"))

> with(Total.P.df, axis(1, at = (1:length(CB3.3e))[
    Month == "Jan"], labels = Year[Month == "Jan"]))
```

These data display seasonal variation, so we need to account for this while testing for trend. Here are the results of the seasonal Kendall test for trend:

```
> kendallSeasonalTrendTest(CB3.3e ~ Month + Year,
    data = Total.P.df)

Results of Hypothesis Test
--------------------------

Null Hypothesis:             All 12 values of tau = 0

Alternative Hypothesis:      The seasonal taus are not all equal
                             (Chi-Square Heterogeneity Test)
                             At least one seasonal tau != 0
                             and all non-zero tau's have the
                             same sign (z Trend Test)

Test Name:                   Seasonal Kendall Test for Trend
                             (with continuity correction)

Estimated Parameter(s):      tau       =  -0.3333333
                             slope     =  -0.2312000
                             intercept = 346.5468681

Estimation Method:           tau:      Weighted Average of
                                       Seasonal Estimates
                             slope:    Hirsch et al.'s
                                       Modification of
                                       Thiel/Sen Estimator
                             intercept: Median of
                                       Seasonal Estimates

Data:                        y       = CB3.3e
                             season = Month
                             year   = Year

Data Source:                 Total.P.df
```

```
Sample Sizes:                          Oct    =   5
                                       Nov    =   5
                                       Dec    =   5
                                       Jan    =   5
                                       Feb    =   5
                                       Mar    =   5
                                       Apr    =   5
                                       May    =   5
                                       Jun    =   5
                                       Jul    =   5
                                       Aug    =   5
                                       Sep    =   5
                                       Total  =  60

Test Statistics:                       Chi-Square (Het)  =   4.480000
                                       z (Trend)          = -2.757716

Test Statistic Parameter:              df = 11

P-values:                              Chi-Square (Het)  = 0.953720141
                                       z (Trend)          = 0.005820666

Confidence Interval for:               slope

Confidence Interval Method:            Gilbert's Modification of
                                       Theil/Sen Method

Confidence Interval Type:              two-sided

Confidence Level:                      95%

Confidence Interval:                   LCL = -0.36162682
                                       UCL = -0.05569319
```

The estimated annual trend is −0.23 mg/year, i.e., a yearly decrease in total phosphorus. The p-value associated with the seasonal Kendall test for trend is $p = 0.006$, indicating this is statistically significant. The two-sided 95 % confidence interval for the trend is [−0.36, −0.06]. The chi-square test for heterogeneity (i.e., is the trend different for different seasons?) yields a p-value of 0.95, so there is no evidence of different amounts of trend within different seasons.

7.6 Summary

- R contains several functions for performing classical statistical hypothesis tests, such as t-tests, analysis of variance, linear regression, nonparametric tests, quality control procedures, and time series analysis (see the R documentation and help files).
- ENVSTATS contains functions for some statistical tests that are not included in R but that are often used in environmental statistics
- Table 7.1 lists functions available in ENVSTATS for performing goodness-of-fit tests.

- Table 7.2 lists functions available in ENVSTATS for performing one-, two- and *k*-sample comparison tests.
- You can test for the presence of serial correlation in a time series or set of residuals from a linear fit using the ENVSTATS function `serialCorrelationTest`.
- Table 7.3 lists the functions available in ENVSTATS for performing nonparametric tests for trend.

Chapter 8

Censored Data

8.1 Introduction

Often in environmental data analysis values are reported simply as being "below detection limit" along with the stated detection limit (e.g., Helsel 2012; Porter et al. 1988; USEPA 1992c, 2001, 2002a, d, 2009; Singh et al. 2002, 2006, 2010b). A sample of data contains *censored observations* if some of the observations are reported only as being below or above some censoring level. Although this results in some loss of information, we can still use data that contain nondetects for graphical and statistical analyses. Statistical methods for dealing with censored data have a long history in the fields of survival analysis and life testing (e.g., Hosmer et al. 2008; Kleinbaum and Klein 2011; Nelson 2004). In this chapter, we will discuss how to create graphs, estimate distribution parameters and quantiles, construct prediction and tolerance intervals, perform goodness-of-fit tests, and compare distributions using censored data. See Helsel (2012) and Millard et al. (2014) for a more in-depth discussion of analyzing environmental censored data.

8.2 Classification of Censored Data

There are four major ways to classify censored data: truncated versus censored, left versus right versus double, single versus multiple (progressive), and censored Type I versus censored Type II (Cohen 1991). Most environmental data sets with nondetect values are either Type I left singly censored or Type I left multiply censored.

A sample of N observations is *left singly censored* (also called singly censored on the left) if c observations are known only to fall below a known censoring level T, while the remaining n ($n = N - c$) uncensored observations falling above T are fully measured and reported.

A sample is *singly censored* (e.g., singly left censored) if there is only one censoring level T. A sample is *multiply censored* or progressively censored (e.g., multiply left censored) if there are several censoring levels T_1, T_2, ..., T_p, where $T_1 < T_2 < ... < T_p$.

A censored sample has been subjected to *Type I censoring* if the censoring level(s) is(are) known in advance, so that given a fixed sample size N, the number of censored observations c (and hence the number of uncensored observations n) is a random outcome. Type I censored samples are sometimes called time-censored samples (Nelson 1982).

8.3 Functions for Censored Data

Table 8.1 lists the functions available in ENVSTATS for analyzing censored data.

Function	Description
ppointsCensored	Compute plotting positions based on censored data
ecdfPlotCensored	Empirical CDF based on censored data
cdfCompareCensored	Compare an empirical CDF to a hypothesized CDF, or compare two CDFs, based on censored data
qqPlotCensored	Q-Q plot based on censored data
boxcoxCensored	Determine an optimal Box-Cox transformation based on censored data
e*abb*Censored	Estimate the parameters of the distribution with the abbreviation *abb* based on censored data, and optionally construct a confidence interval for the parameters
eqnormCensored eqlnormCensored	Estimate the quantiles of the normal or lognormal distribution based on censored data, and optionally construct a confidence interval for a quantile
tolIntNormCensored tolIntLnormCensored	Create a tolerance interval for the normal or lognormal distribution based on censored data
gofTestCensored	Shapiro-Wilk, Shapiro-Francia, and PPCC goodness-of-fit tests for normality for censored data
twoSampleLinear RankTestCensored	Two-sample linear rank test based on censored data

Table 8.1 Functions in ENVSTATS for analyzing censored data

8.4 Graphical Assessment of Censored Data

In Chap. 3 we illustrated several ways of creating graphs for a single variable, including histograms, quantile (empirical cdf) plots, and probability (Q-Q) plots. When you have censored data, creating a histogram is not necessarily straightforward (especially with multiply censored data), but you can create quantile plots and probability plots, as well as determine "optimal" Box-Cox transformations (see Table 8.1).

8.4.1 Quantile (Empirical CDF) Plots for Censored Data

In Chap. 3 we explained that a quantile plot (also called an empirical cumulative distribution function plot or empirical cdf plot) plots the ordered data (the empirical quantiles) on the x-axis versus the estimated cumulative probabilities (or plotting positions) on the y-axis. Various formulas for the plotting positions are given in Millard et al. (2014) and the help file for ecdfPlot. When you have censored data, the formulas for the plotting positions must be modified. For right-censored data, various formulas for the plotting positions are given by Kaplan and Meier

(1958), Nelson (1972), and Michael and Schucany (1986). For left-censored data, formulas for the plotting positions are given by Michael and Schucany (1986), Hirsch and Stedinger (1987), USEPA (2009), and Helsel (2012).

When you have Type I left-censored data with only one censoring level, and all of the uncensored observations are larger than the censoring level, the computation of the plotting positions is straightforward because it is easy to order the uncensored observations. When you have one or more uncensored observations with values less than one or more of the censoring levels, then the computation of the plotting positions becomes a bit trickier. The help file for ppointsCensored in ENVSTATS gives a detailed explanation of the formulas for the plotting positions for censored data.

Silver concentrations (µg/L)								
<0.1	<0.1	0.1	0.1	<0.2	<0.2	<0.2	<0.2	0.2
<0.3	<0.5	0.7	0.8	<1	<1	<1	<1	<1
<1	<1	<1	<1	<1	1	1	1	1.2
1.4	1.5	<2	2	2	2	2	<2.5	2.7
3.2	4.4	<5	<5	<5	<5	5	<6	<10
<10	<10	<10	<10	10	<20	<20	<20	<25
90	560							

Table 8.2 Silver concentrations from an interlab comparison (Helsel and Cohn 1988)

Table 8.2 displays 56 silver concentrations (µg/L) from an interlab comparison that include 34 values below one of 12 detection limits (Helsel and Cohn 1988). These data are stored in the data frame Helsel.Cohn.88.silver.df in EnvStats. Figure 8.1 displays the empirical cdf plot for the silver data. This plot indicates the data are extremely skewed to the right. This is not surprising since looking at the data in the table we see that all of the observations are less than 25 µg/L except for two that are 90 and 560 µg/L. Figure 8.2 displays the quantile plot based on the log-transformed observations. In both of these plots, the upside-down triangles indicate the censoring levels of observations that have been censored. The figures were created with these commands:

```
> with(Helsel.Cohn.88.silver.df,
    ecdfPlotCensored(Ag, Censored,
    xlab = expression(paste("Ag (", mu, "g/L)",
    sep = "")), include.cen = TRUE))

> with(Helsel.Cohn.88.silver.df,
    ecdfPlotCensored(log.Ag, Censored,
    xlab = expression(paste("log [ Ag (", mu, "g/L) ]",
    sep = "")), include.cen = TRUE))
```

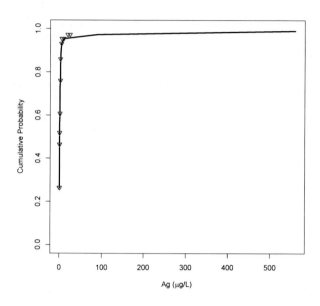

Fig. 8.1 Empirical cdf plot of the silver data using the method of Michael and Schucany (1986)

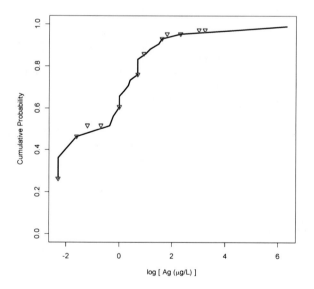

Fig. 8.2 Empirical cdf plot of the log-transformed silver data

8.4.2   Comparing an Empirical and Hypothesized CDF

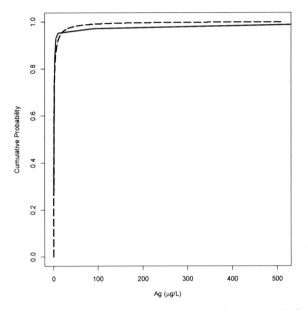

Fig. 8.3 Empirical cdf of the silver data with a fitted lognormal distribution

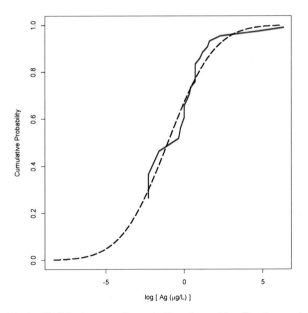

Fig. 8.4 Empirical cdf of the log-transformed silver data with a fitted normal distribution

Figure 8.3 compares the empirical cdf of the silver data with a lognormal cdf, where the parameters for the lognormal distribution are estimated from the data (see the Sect. 8.5 later in this chapter). Figure 8.4 compares the log-transformed silver data with a normal distribution. The figures were created with these commands:

```
> with(Helsel.Cohn.88.silver.df,
    cdfCompareCensored(Ag, Censored, distribution = "lnorm",
    xlab = expression(paste("Ag (", mu, "g/L)", sep = ""))))
```

```
> with(Helsel.Cohn.88.silver.df,
    cdfCompareCensored(log(Ag), Censored, distribution="norm",
    xlab = expression(paste("log [ Ag (", mu, "g/L) ]",
    sep = ""))))
```

The plots appear to show that the lognormal distribution provides an adequate fit to these data but the Shapiro-Francia goodness-of-fit test for lognormality yields a p-value of 0.03 (see the Sect. 8.9.1 later in this chapter).

8.4.3   Comparing Two Empirical CDFs

Table 8.3 displays copper concentrations (μg/L) in shallow groundwater samples from two different geological zones in the San Joaquin Valley, California (Millard and Deverel 1988). The alluvial fan data include four different detection limits and the basin trough data include five different detection limits. In ENVSTATS these data are stored in the data frame Millard.Deverel.88.df. Figure 8.5 compares the empirical cdf of copper concentrations from the alluvial fan zone with those from the basin trough zone. This plot shows that the two distributions are fairly similar in shape and location.

Zone	Copper (μg/L)										
Alluvial fan	<1	<1	<1	<1	1	1	1	1	1	2	2
	2	2	2	2	2	2	2	2	2	2	2
	2	2	2	2	2	2	2	2	3	3	3
	3	3	3	4	4	4	<5	<5	<5	<5	<5
	<5	<5	<5	5	5	5	7	7	7	8	9
	<10	<10	<10	10	11	12	16	<20	<20	20	NA
	NA	NA									
Basin trough	<1	<1	1	1	1	1	1	1	1	<2	<2
	2	2	2	2	3	3	3	3	3	3	3
	3	4	4	4	4	4	<5	<5	<5	<5	<5
	5	6	6	8	9	9	<10	<10	<10	<10	12
	14	<15	15	17	23	NA					

Table 8.3 Copper concentrations in shallow groundwater in two geological zones

```
> attach(Millard.Deverel.88.df)

> Cu.Alluvial <- Cu[Zone == "Alluvial.Fan"]
```

```
> Cu.Alluvial.cen <- Cu.censored[Zone == "Alluvial.Fan"]

> Cu.Basin <- Cu[Zone == "Basin.Trough"]

> Cu.Basin.cen <- Cu.censored[Zone == "Basin.Trough"]

> cdfCompareCensored(Cu.Alluvial, censored = Cu.Alluvial.cen,
    y = Cu.Basin, y.censored = Cu.Basin.cen)
```

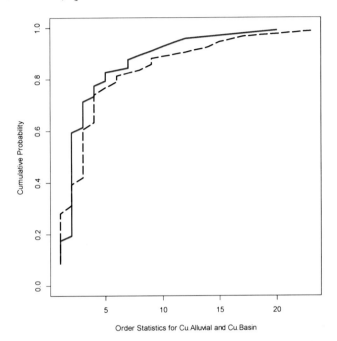

Order Statistics for Cu.Alluvial and Cu.Basin

Fig. 8.5 Empirical CDFs of copper concentrations in the alluvial fan and basin trough zones

8.4.4 Q-Q Plots for Censored Data

In Chap. 3 we explained that a probability plot (also called a quantile-quantile or Q-Q plot) plots the ordered data (the empirical quantiles) on the y-axis versus the corresponding quantiles from the assumed theoretical probability distribution on the x-axis, where the quantiles from the assumed distribution are computed based on the plotting positions. As is the case for empirical cdf plots, when you have censored data, the formulas for the plotting positions must be modified.

Table 8.4 presents artificial TcCB concentrations based on the Reference area TcCB data presented in Sect. 1.11.1 of Chap. 1. For this data set, the concentrations of TcCB less than 0.5 ppb have been recoded as "<0.5," so there are 19 censored observations, 28 uncensored observations, and a total sample size of 47. In ENVSTATS these data are stored in the data frame Modified.TcCB.df.

TcCB concentrations (ppb)							
<0.5	<0.5	<0.5	<0.5	<0.5	<0.5	<0.5	<0.5
<0.5	<0.5	<0.5	<0.5	<0.5	<0.5	<0.5	<0.5
<0.5	<0.5	<0.5	0.5	0.5	0.51	0.52	0.54
0.56	0.56	0.57	0.57	0.6	0.62	0.63	0.67
0.69	0.72	0.74	0.76	0.79	0.81	0.82	0.84
0.89	1.11	1.13	1.14	1.14	1.2	1.33	

Table 8.4 Modified Reference area TcCB concentrations

Figure 8.6 shows the normal Q-Q plot for the log-transformed modified TcCB data, created with this command:

```
> with(Modified.TcCB.df, qqPlotCensored(TcCB, Censored,
    distribution ="lnorm", add.line = TRUE,
    points.col = "blue"))
```

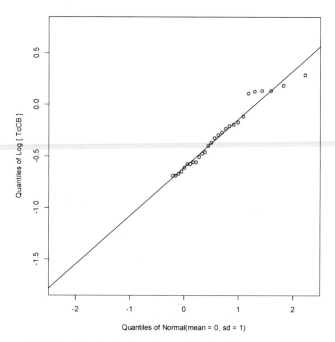

Fig. 8.6 Normal Q-Q plot for the log-transformed singly left-censored TcCB data of Table 8.4 based on the method of Michael and Schucany (1986)

This plot indicates that the lognormal distribution appears to provide an adequate fit to these data. This is not surprising since we already saw in Fig. 1.6 in Sect. 1.11.4 that a lognormal distribution provides a good fit to the original data.

The method that qqPlotCensored uses for computing quantiles is determined by the argument prob.method. The default value of prob.method is "michael-schucany" (Michael and Schucany 1986). Both Helsel (2012)

and USEPA (2009) suggest two other methods: one based on Kaplan-Meier estimates and one based on regression on order statistics (ROS). Setting the value of
prob.method to "kaplan-meier" uses the standard method of Kaplan-
Meier, and when prob.method equals "kaplan-meier with max" the
estimated cumulative probability associated with the maximum value is set to
$(n-0.375)/(n+0.25)$ rather than 1, where n denotes the sample size, i.e., the Blom
plotting position, so that the point associated with the maximum value can be displayed. The ROS method is also discussed below in Sect. 8.5. Other possible
values of prob.method include "nelson" (for right-censored data only) and
"hirsch-stedinger". See the help file for qqPlotCensored for details.

8.4.5 Box-Cox Transformations for Censored Data

In Chap. 3 we discussed using Box-Cox transformations as a way to satisfy
normality assumptions for standard statistical tests, and also sometimes to satisfy
the linear assumption and/or the constant variance in the errors assumption for
a standard linear regression model. We also discussed three possible criteria to
use to decide on the power of the transformation: the probability plot correlation
coefficient (PPCC), the Shapiro-Wilk goodness-of-fit test statistic, and the log-
likelihood function. This idea can be extended to the case of censored data (e.g.,
Shumway et al. 1989). See the help file for boxcoxCensored for details.

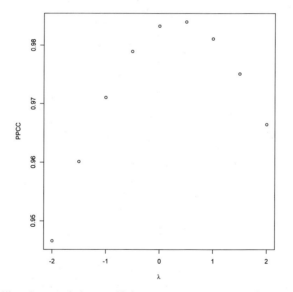

Fig. 8.7 Probability plot correlation coefficient versus Box-Cox transform power (λ) for the
singly censored TcCB data of Table 8.4

Figure 8.7 displays a plot of the probability plot correlation coefficient versus various values of the transform power λ for the singly censored TcCB data shown in Table 8.4, created with these commands:

```
> boxcox.list <- with(Modified.TcCB.df,
    boxcoxCensored(TcCB, Censored))

> plot(boxcox.list)
```

For these data, the PPCC reaches its maximum between about λ = 0 (log transformation) and λ = 0.5 (square-root transformation). We saw a similar pattern for the original Reference area TcCB data in Fig. 3.8 in Sect. 3.8, although in that figure the maximum appeared at about λ = 0.

8.5   Estimating Distribution Parameters

In Chap. 5 we illustrated how to use ENVSTATS to estimate distribution parameters and quantiles, and also create confidence intervals for these quantities. Methods for estimating parameters include maximum likelihood, method of moments, and minimum variance unbiased. It is fairly straightforward to extend maximum likelihood estimation to the case of censored data (e.g., Cohen 1991; Schneider 1986). More recently, researchers in the environmental field have proposed alternative methods of computing estimates and confidence intervals in addition to the classical ones such as maximum likelihood estimation. See Helsel (2012) and Millard et al. (2014) for details.

8.5.1   The Normal and Lognormal Distribution

Estimating Parameters

The function `enormCensored` estimates the mean and standard deviation of a normal distribution based on Type I censored data, and optionally constructs a confidence interval for the mean. The function `elnormCensored` does the same thing assuming a lognormal distribution and estimating the mean and standard deviation on the log-scale. These functions both take an argument called `method` that determines what estimation method is used. Table 8.5 lists the estimation methods available along with references. The half censoring level method (`method="half.cen.level"`) in which left-censored observations are set to half the value of the censoring level associated with that observation is included only for historical reasons and should never be used (e.g., Helsel 2012), except perhaps in simulations to show how poorly it performs!

Chapter 15 of USEPA (2009) gives several examples of estimating the mean and standard deviation of a lognormal distribution on the log-scale using manganese concentrations (ppb) in groundwater at five background wells. In EnvStats these data are stored in the data frame `EPA.09.Ex.15.1.manganese.df`.

```
> EPA.09.Ex.15.1.manganese.df

  Sample  Well Manganese.Orig.ppb Manganese.ppb Censored
1      1 Well.1                <5           5.0     TRUE
2      2 Well.1              12.1          12.1    FALSE
...
24     4 Well.5               8.4           8.4    FALSE
25     5 Well.5                <2           2.0     TRUE

> longToWide(EPA.09.Ex.15.1.manganese.df,
    "Manganese.Orig.ppb", "Sample", "Well",
    paste.row.name = TRUE)

         Well.1 Well.2 Well.3 Well.4 Well.5
Sample.1     <5     <5     <5    6.3   17.9
Sample.2   12.1    7.7    5.3   11.9   22.7
Sample.3   16.9   53.6   12.6     10    3.3
Sample.4   21.6    9.5  106.3     <2    8.4
Sample.5     <2   45.9   34.5   77.2     <2
```

Value of method	Common name	References
mle	Maximum likelihood estimator (MLE)	Cohen (1959, 1991)
bcmle	Bias-corrected MLE	Saw (1961b), Schneider (1986), Haas and Scheff (1990), Bain and Engelhardt (1991)
qq.reg	Q-Q regression. Also called probability plot method, and regression on order statistics (ROS).	Nelson (1982), Gilbert (1987), Hass and Scheff (1990), Travis and Land (1990), Helsel and Hirsch (1992)
qq.reg.w.cen.level	Q-Q regression with censoring level	El-Shaarawi (1989)
impute.w.qq.reg	Imputation with Q-Q Regression. Also called robust regression on order statistics (Robust ROS)	Hashimoto and Trussell (1983), Gilliom and Helsel (1986), El-Shaarawi (1989), Helsel (2012)
impute.w.qq.reg.w.cen.level	Impute with Q-Q regression with censoring level	El-Shaarawi (1989)
impute.w.mle	Impute with MLE	El-Shaarawi (1989)
iterative.impute.w.qq.reg	Iterative imputation with Q-Q regression	Gleit (1985)
m.est	Robust M estimation	Korn and Tyler (2001)
half.cen.level	Half censoring level	Gleit (1985), Haas and Scheff (1990), El-Shaarawi and Esterby (1992)

Table 8.5 Available methods for estimating the parameters of a normal distribution based on censored data using the functions enormCensored or elnormCensored

Here we will estimate the mean and standard deviation using the MLE, Q-Q regression (also called parametric regression on order statistics or ROS; e.g., USEPA 2009 and Helsel 2012), and imputation with Q-Q regression (also called robust ROS). The command to estimate the parameters using the MLE method is:

```
> with(EPA.09.Ex.15.1.manganese.df,
    elnormCensored(Manganese.ppb, Censored))

Results of Distribution Parameter Estimation
Based on Type I Censored Data
--------------------------------------------

Assumed Distribution:            Lognormal

Censoring Side:                  left

Censoring Level(s):              2 5

Estimated Parameter(s):          meanlog = 2.215905
                                 sdlog   = 1.356291

Estimation Method:               MLE

Data:                            Manganese.ppb

Censoring Variable:              Censored

Sample Size:                     25

Percent Censored:                24%
```

The commands to estimate the parameters based on Q-Q regression (ROS) and imputation with Q-Q regression (robust ROS) are, respectively:

```
> with(EPA.09.Ex.15.1.manganese.df,
    elnormCensored(Manganese.ppb, Censored,
    method = "qq.reg"))
```

```
> with(EPA.09.Ex.15.1.manganese.df,
    elnormCensored(Manganese.ppb, Censored,
    method = "impute.w.qq.reg"))
```

The estimated mean and standard deviation are 2.29 and 1.28 ppb based on Q-Q regression, and 2.30 and 1.24 ppb based on imputation with Q-Q regression.

As explained in Sect. 8.4.4, the method used to estimate quantiles for a Q-Q plot is determined by the argument prob.method. For the functions enormCensored and elnormCensored, for any estimation method that involves Q-Q regression, the default value of prob.method is "hirsch-stedinger" and the default value for the plotting position constant is plot.pos.con=0.375. Both Helsel (2012) and USEPA (2009) also use the Hirsch-Stedinger probability method but set the plotting position constant to 0. In this case the estimated mean and standard deviation are 2.28 and 1.26 ppb.

```
> with(EPA.09.Ex.15.1.manganese.df,
    elnormCensored(Manganese.ppb, Censored,
    method = "impute.w.qq.reg", plot.pos.con = 0))
```

Computing a Confidence Interval for the Mean

The functions enormCensored and elnormCensored take an argument called ci.method that determines what method is used to construct a confidence interval for the mean. Table 8.6 lists the available methods along with references.

Value of ci.method	Common name	References
profile.likelihood	Profile likelihood	Venzon and Moolgavkar (1988)
normal.approx	Normal approximation	Cohen (1959, 1991)
normal.approx.w.cov	Normal approximation using variance-covariance of parameter estimates	Schneider (1986)
bootstrap	Bootstrap	Efron and Tibshirani (1993)
gpq	Generalized pivotal quantity	Schmee et al. (1985), Krishnamoorthy and Mathew (2009)

Table 8.6 Available methods for constructing a confidence interval for the mean of a normal distribution based on censored data

Here is the command to produce a 95 % confidence interval for the mean of the manganese data using the MLE method to estimate the parameters and the profile likelihood method to produce the confidence interval:

```
> with(EPA.09.Ex.15.1.manganese.df,
    elnormCensored(Manganese.ppb, Censored, ci = TRUE))

Results of Distribution Parameter Estimation
Based on Type I Censored Data
--------------------------------------------

Assumed Distribution:            Lognormal

Censoring Side:                  left

Censoring Level(s):              2 5
```

```
Estimated Parameter(s):        meanlog = 2.215905
                               sdlog   = 1.356291

Estimation Method:             MLE

Data:                          Manganese.ppb

Censoring Variable:            Censored

Sample Size:                   25

Percent Censored:              24%

Confidence Interval for:       meanlog

Confidence Interval Method:    Profile Likelihood

Confidence Interval Type:      two-sided

Confidence Level:              95%

Confidence Interval:           LCL = 1.595062
                               UCL = 2.771197
```

8.5.2   The Lognormal Distribution (Original Scale)

Value of method	Common name	References
mle	Maximum likelihood estimator (MLE)	Cohen (1959, 1991)
bcmle	Bias-corrected MLE	El-Shaarawi (1989)
qmvue	Quasi-minimum variance un-biased estimator (QMVUE)	Gilliom and Helsel (1986), Newman et al. (1989), Cohn et al. (1989)
impute.w.qq.reg	Imputation with Q-Q regres-sion. Also called robust re-gression on order statistics (robust ROS)	Hashimoto and Trussell (1983), Gilliom and Helsel (1986), El-Shaarawi (1989), Helsel (2012)
impute.w.qq.reg.w. cen.level	Impute with Q-Q regression with censoring level	El-Shaarawi (1989)
impute.w.mle	Impute with MLE	El-Shaarawi (1989)
half.cen.level	Half censoring level	Gleit (1985), Haas and Scheff (1990), El-Shaarawi and Esterby (1992)

Table 8.7 Available methods for estimating the parameters of a lognormal distribution based on censored data using the function elnormAltCensored

The function `elnormAltCensored` estimates the mean and coefficient of variation of a lognormal distribution on the original scale. Table 8.7 shows the available estimation methods (as for the normal distribution, the half censoring level method should only be used in simulations to show how poorly it performs), and Table 8.8 shows the methods available for constructing a confidence interval for the mean on the original scale.

Value of `ci.method`	Common name	References
`profile.likelihood`	Profile likelihood	Venzon and Moolgavkar (1988)
`cox`	Extension of Cox's method to censored data	El-Shaarawi (1989)
`delta`	Delta method	Shumway et al. (1989)
`normal.approx`	Normal approximation	Cohen (1959, 1991)
`bootstrap`	Bootstrap	Efron and Tibshirani (1993)

Table 8.8 Available methods for constructing a confidence interval for the mean of a lognormal distribution based on censored data

In Sect. 5.2.2 we estimated the mean TcCB concentration in the Reference area as 0.6 ppb and the CV as 0.49, assuming the data come from a lognormal distribution. We also computed a two-sided 95% confidence intervals for the mean as [0.52, 0.70]. There are 47 uncensored observations in this data set. The modified TcCB data shown in Table 8.4 include 19 censored observations. Using these data, the estimated mean and CV are 0.61 and 0.46 based on the maximum likelihood method, and the two-sided 95 % confidence interval for the mean is [0.53, 0.70] based on the profile likelihood method.

```
> with(Modified.TcCB.df,
    elnormAltCensored(TcCB, Censored, ci = TRUE))

Results of Distribution Parameter Estimation
Based on Type I Censored Data
---------------------------------------------


Assumed Distribution:            Lognormal

Censoring Side:                  left

Censoring Level(s):              0.5

Estimated Parameter(s):          mean = 0.6082981
                                 cv   = 0.4645955

Estimation Method:               MLE
```

```
Data:                              TcCB

Censoring Variable:                Censored

Sample Size:                       47

Percent Censored:                  40.42553%

Confidence Interval for:           mean

Confidence Interval Method:        Profile Likelihood

Confidence Interval Type:          two-sided

Confidence Level:                  95%

Confidence Interval:               LCL = 0.5292601
                                   UCL = 0.7032881
```

8.5.3   The Gamma Distribution

The function egammaCensored estimates the shape and scale parameters of a gamma distribution based on censored data, while the function egammaAltCensored estimates the mean and coefficient of variation. Currently the only available method of estimation for these functions is maximum likelihood estimation. Methods for constructing a confidence interval for the mean include the normal approximation, bootstrap, and profile likelihood. Here is the command to estimate the mean and CV and create a 95% confidence interval for the mean using the modified TcCB data shown in Table 8.4:

```
> with(Modified.TcCB.df,
    egammaAltCensored(TcCB, Censored, ci = TRUE))
```

```
Results of Distribution Parameter Estimation
Based on Type I Censored Data
-----------------------------------------------

Assumed Distribution:              Gamma

Censoring Side:                    left

Censoring Level(s):                0.5
```

```
Estimated Parameter(s):          mean = 0.5981132
                                 cv   = 0.4739765

Estimation Method:               MLE

Data:                            TcCB

Censoring Variable:              Censored

Sample Size:                     47

Percent Censored:                40.42553%

Confidence Interval for:         mean

Confidence Interval Method:      Profile Likelihood

Confidence Interval Type:        two-sided

Confidence Level:                95%

Confidence Interval:             LCL = 0.5127789
                                 UCL = 0.6895852
```

In this case, the estimated mean and CV and the confidence interval for the mean are nearly identical to the values we saw in the previous section when we assumed a lognormal distribution.

8.5.4   Estimating the Mean Nonparametrically

Formulas for the mean and the variance of the mean based on right-censored data using the Kaplan-Meier estimate of the survival function are well known from the survival analysis literature (e.g., Lee and Wang 2003). Helsel (2012) discusses adapting these formulas for left-censored data as well in order to estimate the mean nonparametrically and also to construct a confidence interval for the mean based on a normal approximation. (He also notes you should not use this method for left-censored data if there is only one censoring level since it is then equivalent to substituting the detection limit for the censored observations.) You can implement these methods in **EnvStats** using the enparCensored function.

Following Example 15-1 in USEPA (2009, p. 15–10), we'll use the manganese data introduced in Sect. 8.5.1 to nonparametrically estimate the mean on the log-scale and create a 95 % confidence interval for the mean:

```
> with(EPA.09.Ex.15.1.manganese.df,
    enparCensored(log(Manganese.ppb), Censored, ci = TRUE))
```

Results of Distribution Parameter Estimation
Based on Type I Censored Data

Assumed Distribution: None

Censoring Side: left

Censoring Level(s):   0.6931472 1.6094379

Estimated Parameter(s):   mean   = 2.3092890
                                   sd   = 1.1816102
                                   se.mean = 0.1682862

Estimation Method: Kaplan-Meier

Data: log(Manganese.ppb)

Censoring Variable: Censored

Sample Size: 25

Percent Censored: 24%

Confidence Interval for: mean

Confidence Interval Method: Normal Approximation

Confidence Interval Type: two-sided

Confidence Level: 95%

Confidence Interval: LCL = 1.979454
                                   UCL = 2.639124
```

The estimated mean and standard deviation, as well as the confidence interval for the mean, are not too different from the results based on assuming a lognormal distribution that were shown in Sect. 8.5.1  Of course, if you are estimating the mean nonparametrically, there is no reason to use a log transformation to attempt to induce normality; here is the command to estimate the mean and standard deviation on the original scale:
```

```
> with(EPA.09.Ex.15.1.manganese.df,
    enparCensored(Manganese.ppb, Censored, ci = TRUE))
```

Results of Distribution Parameter Estimation
Based on Type I Censored Data

Assumed Distribution: None

Censoring Side: left

Censoring Level(s):   2 5

Estimated Parameter(s):   mean   = 19.867000
                                 sd   = 25.317737
                                 se.mean = 4.689888

Estimation Method: Kaplan-Meier

Data: Manganese.ppb

Censoring Variable: Censored

Sample Size: 25

Percent Censored: 24%

Confidence Interval for: mean

Confidence Interval Method: Normal Approximation

Confidence Interval Type: two-sided

Confidence Level: 95%

Confidence Interval: LCL = 10.67499
                                 UCL = 29.05901

8.6   Estimating Distribution Quantiles

In Chap. 5 we illustrated how to estimate and construct confidence intervals for population quantiles or percentiles, both parametrically and nonparametrically. In this section we will discuss how to do this with censored data.

8.6.1   Parametric Estimates of Quantiles

A parametric estimate of a quantile is a function of the estimated distribution parameters (e.g., mean and standard deviation). The same is true for a confidence interval for the quantile. To estimate a quantile in the presence of censored observations, you can simply use the same formula for estimating the quantile as for complete data, but estimate the distribution parameters using formulas that are appropriate for censored data (e.g., USEPA 2009; Helsel 2012). If you use maximum likelihood estimation for the distribution parameters, then the resulting estimate of the quantile is also a maximum likelihood estimate. Estimates of quantiles based on other kinds of estimates of the distribution parameters (e.g., ROS, robust ROS, Kaplan-Meier, etc.) should have similar properties, but this is an area that requires further research. Also, it is not clear how well this method works for producing accurate confidence intervals for a quantile. An alternative method for constructing a parametric confidence interval for a quantile that has been shown to perform fairly well is based on generalized pivotal quantities or GPQs (Krishnamoorthy and Mathew 2009).

The Normal and Lognormal Distribution

The EnvStats functions eqnormCensored and eqlnormCensored allow you to estimate quantiles and construct intervals for them assuming a normal or lognormal distribution, respectively. You can construct confidence intervals by using the standard formulas for complete data and substituting in estimates of the mean and standard deviation based on an appropriate formula for censored data, or you can use the GPQ method.

Using the Reference area TcCB concentrations of Table 1.1 in Sect. 1.11.1, the estimated 90th percentile is 0.98 ppb, assuming these data come from a lognormal distribution. Also, the two-sided 95 % confidence interval for the 90th percentile is [0.84, 1.22]. If instead we use the modified TcCB concentrations shown in Table 8.4, the estimated 90th percentile is 0.97 using the MLE, and the two-sided 95 % confidence interval for the 90th percentile is [0.84, 1.19] using the standard formula for complete data but plugging in the MLEs for the mean and standard deviation, so in this case censoring does not have a large effect on the estimate or confidence interval.

```
> with(Modified.TcCB.df,
    eqlnormCensored(TcCB, Censored, p = 0.9, ci = TRUE))

Results of Distribution Parameter Estimation
Based on Type I Censored Data
--------------------------------------------------


Assumed Distribution:            Lognormal

Censoring Side:                  left
```

```
Censoring Level(s):                 0.5

Estimated Parameter(s):             meanlog = -0.5948115
                                    sdlog   =  0.4420888

Estimation Method:                  MLE

Estimated Quantile(s):              90'th %ile = 0.9721435

Quantile Estimation Method:         Quantile(s) Based on
                                    MLE Estimators

Data:                               TcCB

Censoring Variable:                 Censored

Sample Size:                        47

Percent Censored:                   40.42553%

Confidence Interval for:            90'th %ile

Assumed Sample Size:                47

Confidence Interval Method:         Exact for
                                    Complete Data

Confidence Interval Type:           two-sided

Confidence Level:                   95%

Confidence Interval:                LCL = 0.8362336
                                    UCL = 1.1911067
```

To estimate the quantile based on distribution parameters estimated using imputation with Q-Q regression (robust ROS), and to compute the confidence interval using the method of generalized pivotal quantities, type this command:

```
> with(Modified.TcCB.df, eqlnormCensored(TcCB, Censored,
    p = 0.9, method = "impute.w.qq.reg", ci = TRUE,
    ci.method = "gpq", seed = 47))
```

```
Results of Distribution Parameter Estimation
Based on Type I Censored Data
-----------------------------------------------

Assumed Distribution:            Lognormal

Censoring Side:                  left

Censoring Level(s):              0.5

Estimated Parameter(s):          meanlog = -0.6098130
                                 sdlog   =  0.4604909

Estimation Method:               Imputation with
                                 Q-Q Regression (ROS)

Estimated Quantile(s):           90'th %ile = 0.980522

Quantile Estimation Method:      Quantile(s) Based on
                                 Imputation with
                                 Q-Q Regression (ROS) Estimators

Data:                            TcCB

Censoring Variable:              Censored

Sample Size:                     47

Percent Censored:                40.42553%

Confidence Interval for:         90'th %ile

Confidence Interval Method:      Generalized Pivotal Quantity

Number of Monte Carlos:          1000

Confidence Interval Type:        two-sided

Confidence Level:                95%

Confidence Interval:             LCL = 0.8462989
                                 UCL = 1.2525365
```

The argument seed lets you set the seed for the random number generator used in the Monte Carlo trials so you can reproduce the exact same results shown here. Note that by default the GPQ method of constructing the confidence interval uses only 1,000 Monte Carlo trials, whereas Krishnamoorthy and Mathew (2009) suggest using 10,000 Monte Carlo trials.

Other Distributions

All of the available functions for estimating quantiles based on complete data (see Sect. 5.3) accept objects that are the result of distribution parameter estimation, so you can estimate distribution parameters based on censored data and supply these results to functions that estimate quantiles based on complete data. (However, as previously stated, it is not clear how well this method performs.) For example, here is the command to estimate the 90th percentile using the modified TcCB concentrations shown in Table 8.4 and assuming the data come from a gamma distribution:

```
> egamma.list <- with(Modified.TcCB.df,
    egammaCensored(TcCB, Censored))

> eqgamma(egamma.list, p = 0.9)

Results of Distribution Parameter Estimation
Based on Type I Censored Data
--------------------------------------------


Assumed Distribution:              Gamma

Censoring Side:                    left

Censoring Level(s):                0.5

Estimated Parameter(s):            shape = 4.4512947
                                   scale = 0.1343684

Estimation Method:                 MLE

Estimated Quantile(s):             90'th %ile = 0.9779156

Quantile Estimation Method:        Quantile(s) Based on
                                   MLE Estimators

Data:                              TcCB

Censoring Variable:                Censored
```

```
Sample Size:                          47

Percent Censored:                     40.42553%
```

8.6.2   Nonparametric Estimates of Quantiles

We saw in Chap. 5 that nonparametric estimates and confidence intervals for population percentiles are simply functions of the ordered observations. Thus, for left censored data, you can still estimate quantiles and create one-sided upper confidence intervals as long as there are enough uncensored observations that can be ordered in a logical way, just as we did in Sect. 5.3.4 using the nitrate data.

8.7   Prediction Intervals

In Chap. 6 we illustrated how to construct parametric and nonparametric prediction intervals. In this section we will discuss how to do this with censored data.

8.7.1   Parametric Prediction Intervals

Just as for parametric estimates of a quantile, a parametric prediction interval is a function of the estimated distribution parameters. To construct a parametric prediction interval in the presence of censored observations, you can simply use the same formula as for complete data, but estimate the distribution parameters using formulas that are appropriate for censored data (e.g., USEPA 2009; Helsel 2012). Again, it is not clear how well this method works for producing accurate prediction intervals; this is an area that requires further research.

All of the available functions for constructing prediction intervals based on complete data (see Sect. 6.2) accept objects that are the result of distribution parameter estimation, so you can estimate distribution parameters based on censored data and supply these results to functions that compute prediction intervals based on complete data. For example, in Sect. 6.2.4, we constructed a 95 % upper nonparametric prediction interval for the next $k = 4$ future monthly trichloroethylene concentrations (ppb) in groundwater at a compliance well using the data from three background wells shown in Table 6.3. Here are the commands to produce a parametric version of this prediction interval assuming a normal distribution:

```
> enorm.list <- with(EPA.09.Ex.18.3.TCE.df,
    enormCensored(TCE.ppb[Well.type == "Background"],
    Censored[Well.type == "Background"]))

> predIntNorm(enorm.list, k = 4, pi.type = "upper",
    conf.level = 0.95, method = "exact")
```

```
Results of Distribution Parameter Estimation
Based on Type I Censored Data
--------------------------------------------

Assumed Distribution:            Normal

Censoring Side:                  left

Censoring Level(s):              5

Estimated Parameter(s):          mean = 5.297871
                                 sd   = 3.981489

Estimation Method:               MLE

Data:                   TCE.ppb[Well.type == "Background"]

Censoring Variable:     Censored[Well.type == "Background"]

Sample Size:                     18

Percent Censored:                50%

Assumed Sample Size:             18

Prediction Interval Method:      exact

Prediction Interval Type:        upper

Confidence Level:                95%

Number of Future Observations:   4

Prediction Interval:             LPL =    -Inf
                                 UPL = 15.2453
```

Note that although the prediction interval method in the returned object has the value "exact", it is not in fact exact because the prediction interval is based on censored, rather than complete, data. Because half of the observations are censored, it is difficult to determine with say the function qqPlotCensored whether these data appear to fit a normal, lognormal, or some other distribution.

8.7.2   Nonparametric Prediction Intervals

We saw in Chap. 6 that nonparametric prediction intervals are simply functions of the ordered observations. Thus, for left censored data, you can still create nonparametric prediction intervals as long as there are enough uncensored observations that can be ordered in a logical way, just as we did in Sect. 6.2.4 with the trichloroethylene and xylene data and in Sect. 6.3.4 with the mercury data.

8.8   Tolerance Intervals

In Chap. 6 we illustrated how to construct parametric and nonparametric tolerance intervals. In this section we will discuss how to do this with censored data

8.8.1   Parametric Tolerance Intervals

Just as for parametric estimates of a quantile, a parametric tolerance interval is a function of the estimated distribution parameters. To construct a parametric tolerance interval in the presence of censored observations, you can simply use the same formula as for complete data, but estimate the distribution parameters using formulas that are appropriate for censored data (e.g., USEPA 2009; Helsel 2012). Again, it is not clear how well this method works for producing accurate tolerance intervals; this is an area that requires further research. An alternative method for constructing parametric tolerance intervals that has been shown to perform fairly well is based on generalized pivotal quantities or GPQs (Krishnamoorthy and Mathew 2009).

The functions tolIntNormCensored and tolIntLnormCensored allow you to construct tolerance intervals assuming a normal or lognormal distribution, respectively. You can construct tolerance intervals by using the standard formulas for complete data and substituting in estimates of the mean and standard deviation based on an appropriate formula for censored data, or you can use the GPQ method.

In Chap. 6, Sect. 6.4.2, we constructed a one-sided upper tolerance limit based on using the Reference area TcCB data presented in Sect. 1.11.1 of Chap. 1 and assuming a lognormal distribution. The tolerance limit was a 95 % β-content tolerance limit with associated confidence level of 95 % and had the value 1.42 ppb. For the singly censored data set shown in Table 8.4, this limit becomes 1.38 ppb based on using MLEs to estimate the mean and standard deviation and using the formula for the tolerance interval that assumes complete data. In this case there is very little difference between the limits based on the complete data versus those based on the censored data. Here is the command to construct the tolerance limit using the standard formula for complete data but plugging in the MLEs for the mean and standard deviation:

```
> with(Modified.TcCB.df,
    tolIntLnormCensored(TcCB, Censored, ti.type = "upper"))
```

```
Results of Distribution Parameter Estimation
Based on Type I Censored Data
--------------------------------------------

Assumed Distribution:              Lognormal

Censoring Side:                    left

Censoring Level(s):                -0.6931472

Estimated Parameter(s):            meanlog = -0.5948115
                                   sdlog   =  0.4420888

Estimation Method:                 MLE

Data:                              TcCB

Censoring Variable:                censored

Sample Size:                       47

Percent Censored:                  40.42553%

Assumed Sample Size:               47

Tolerance Interval Coverage:       95%

Coverage Type:                     content

Tolerance Interval Method:         Exact for
                                   Complete Data

Tolerance Interval Type:           upper

Confidence Level:                  95%

Tolerance Interval:                LTL = 0.000000
                                   UTL = 1.384159
```

To construct the tolerance interval based on distribution parameters estimated using imputation with Q-Q regression (robust ROS), and using the method of generalized pivotal quantities, type this command:

```
> with(Modified.TcCB.df, tolIntLnormCensored(TcCB, Censored,
    ti.type = "upper", method = "impute.w.qq.reg",
    ti.method = "gpq", seed = 47))
```

Results of Distribution Parameter Estimation
Based on Type I Censored Data

Assumed Distribution: Lognormal

Censoring Side: left

Censoring Level(s):   -0.6931472

Estimated Parameter(s):   meanlog = -0.6098130
                                   sdlog   = 0.4604909

Estimation Method: Imputation with
                                   Q-Q Regression (ROS)

Data: TcCB

Censoring Variable: censored

Sample Size: 47

Percent Censored: 40.42553%

Tolerance Interval Coverage: 95%

Coverage Type: content

Tolerance Interval Method: Generalized Pivotal Quantity

Number of Monte Carlos: 1000

Tolerance Interval Type: upper

Confidence Level: 95%

Tolerance Interval: LTL = 0.000000
                                   UTL = 1.492929

8.8.2 Nonparametric Tolerance Intervals

We saw in Chap. 6 that nonparametric tolerance intervals are simply functions of the ordered observations. Thus, for left censored data, you can still create nonparametric tolerance intervals as long as there are enough uncensored observations that can be ordered in a logical way, just as we did in Sect. 6.4.4 with the copper data.

8.9 Hypothesis Tests

In this section we discuss how to perform hypothesis tests in the presence of censored data.

8.9.1 Goodness-of-Fit Tests

Royston (1993) extended both the Shapiro-Francia and Shapiro-Wilk goodness-of-fit tests to the case of singly censored data. He also provides a method of computing p-values for these statistics based on tables given in Verrill and Johnson (1988). Although Verrill and Johnson (1988) produced their tables based on Type II censoring, Royston's (1993) approximation to the p-value of these tests should be fairly accurate for Type I censored data as well. The Shapiro-Francia and PPCC tests are also easily extendible to the case of multiply censored data, but it is not known how well Royston's method of computing p-values works in this case. The ENVSTATS function gofTestCensored performs goodness-of-fit tests for censored data using Royson's method; see the help file for details.

We noted in Chap. 7 that goodness-of-fit tests are of limited value for small sample sizes because there is usually not enough information to distinguish between different kinds of distributions. This also holds true even for moderate sample sizes if you have data with a moderate amount of censored observations, as the next example shows.

In Sect. 7.2.1 we showed that goodness-of-fit tests for the Reference area TcCB data indicated that the normal distribution was not appropriate, but that a lognormal distribution appeared to adequately model the data. In this example we will perform the same tests but use the modified TcCB data of Table 8.4. The Shapiro-Wilk test for normality yields a p-value of 0.35 and the test for lognormality yields a p-value of 0.43. Figures 8.8 and 8.9 show companion plots for the tests for normality and lognormality, respectively. Compare these figures to Figs. 7.1 and 7.2. We see that unlike the case with complete data, here censoring 40 % of the observations leaves us unable to distinguish between a normal and lognormal distribution.

```
> sw.list.norm <- with(Modified.TcCB.df,
    gofTestCensored(TcCB, Censored))

> sw.list.norm
```

```
Results of Goodness-of-Fit Test
Based on Type I Censored Data
-------------------------------
```

Test Method: Shapiro-Wilk GOF
                                    (Singly Censored Data)

Hypothesized Distribution: Normal

Censoring Side: left

Censoring Level(s):   0.5

Estimated Parameter(s):   mean = 0.5580721
                                    sd   = 0.3371722

Estimation Method: MLE

Data: TcCB

Censoring Variable: Censored

Sample Size: 47

Percent Censored: 40.4%

Test Statistic: W = 0.9625386

Test Statistic Parameters: N   = 47.0000000
                                    DELTA = 0.4042553

P-value: 0.3469034

Alternative Hypothesis: True cdf does not equal the
                                    Normal Distribution.

```
> sw.list.lnorm <- with(Modified.TcCB.df,
    gofTestCensored(TcCB, Censored, dist = "lnorm"))
> sw.list.lnorm
```

```
Results of Goodness-of-Fit Test
Based on Type I Censored Data
-------------------------------
```

Test Method: Shapiro-Wilk GOF
                                (Singly Censored Data)

Hypothesized Distribution: Lognormal

Censoring Side: left

Censoring Level(s):   0.5

Estimated Parameter(s):   meanlog = -0.5948115
                                sdlog   = 0.4420888

Estimation Method: MLE

Data: TcCB

Censoring Variable: Censored

Sample Size: 47

Percent Censored: 40.4%

Test Statistic: W = 0.9667591

Test Statistic Parameters: N   = 47.0000000
                                DELTA = 0.4042553

P-value: 0.4298383

Alternative Hypothesis: True cdf does not equal the
                                Lognormal Distribution.

To plot the results of these tests as shown in Figs. 8.8 and 8.9, type these commands.

> plot(sw.list.norm)

> plot(sw.list.lnorm)

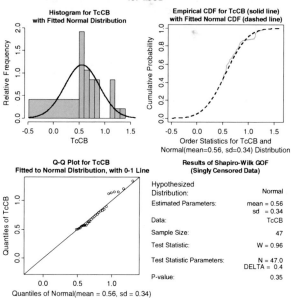

Fig. 8.8 Companion plots for the Shapiro-Wilk test for normality for the singly censored Reference area TcCB data

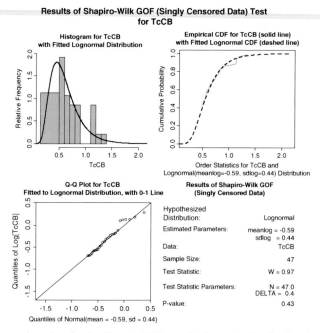

Fig. 8.9 Companion plots for the Shapiro-Wilk test for lognormality for the singly censored Reference area TcCB data

8.9.2   Nonparametric Tests to Compare Two Groups

In Sect. 7.3 of Chap. 7 we discussed various hypothesis tests to compare locations (central tendency) between two groups, including the Wilcoxon rank sum test, other linear rank tests, and the quantile test. In the presence of censored observations, you can still use a linear rank test (e.g., the Wilcoxon rank sum test) or quantile test as long as there are enough uncensored observations that can be ordered in a logical way. For example, if both samples are singly censored with the same censoring level and all uncensored observations are greater than the censoring level, then all censored observations receive the lowest ranks and are considered tied observations. Actually you can use linear rank tests even with multiply censored data as well.

We stated in Chap. 7 that the Wilcoxon rank sum test is a particular kind of linear rank test. Several authors have proposed extensions of the Wilcoxon rank sum test to the case of singly or multiply censored data, mainly in the context of survival analysis (e.g., Breslow 1970; Cox 1972; Gehan 1965; Mantel 1966; Peto and Peto 1972; Prentice 1978). Prentice (1978) showed how all of these proposed tests are extensions of a linear rank test to the case of censored observations. As for the case of complete data, different linear rank tests use different score functions, and some may be better than others at detecting a small shift in location, depending upon the true underlying distribution.

Prentice and Marek (1979), Latta (1981), and Millard and Deverel (1988) studied the behavior of several linear rank tests for censored data. For details, see the help file for twoSampleLinearRankTestCensored.

In Sect. 8.4.3 we compared the empirical cumulative distribution functions of copper concentrations in the alluvial fan and basin trough zones (see Table 8.3 and Fig. 8.5). The plot indicates the distributions of concentrations are fairly similar. The two-sample linear rank test based on normal scores and a hypergeometric variance yields a p-value of 0.2, indicating no significant difference. To perform the two-sample linear rank test to compare copper concentrations, type these commands.

```
> attach(Millard.Deverel.88.df)

> Cu.AF <- Cu[Zone=="Alluvial.Fan"]

> Cu.AF.cen <- Cu.censored[Zone=="Alluvial.Fan"]

> Cu.BT <- Cu[Zone=="Basin.Trough"]

> Cu.BT.cen <- Cu.censored[Zone=="Basin.Trough"]

> twoSampleLinearRankTestCensored(Cu.AF, Cu.AF.cen,
    Cu.BT, Cu.BT.cen, test = "normal.scores.2",
    var = "hypergeometric")
```

```
Results of Hypothesis Test
Based on Censored Data
--------------------------
```

Null Hypothesis: Fy(t) = Fx(t)

Alternative Hypothesis: Fy(t) != Fx(t) for at least
one t

Test Name: Two-Sample Linear Rank Test:
                                    Normal Scores Test Using
                                    Prentice Survival Estimator
                             Based on Prentice and Marek (1979)
                                    with Hypergeometric Variance

Censoring Side: left

Censoring Level(s)   x = 1 5 10 20
                                    y = 1 2 5 10 15

Data: x = Cu.AF
                                    y = Cu.BT

Censoring Variable: x = Cu.AF.cen
                                    y = Cu.BT.cen

Number NA/NaN/Inf's Removed: x = 3
                                    y = 1

Sample Sizes: nx = 65
                                    ny = 49

Percent Censored: x = 26.2%
                                    y = 28.6%

Test Statistics: nu   = -5.119320
                                    var.nu = 16.020363
                                    z   = -1.279016

P-value: 0.2008913

8.10 Summary

- A sample of data contains *censored observations* if some of the observations are reported only as being below or above some censoring level.
- Environmental data often contain values reported as "less-than-detection-limit".
- Table 8.1 lists functions available in ENVSTATS for analyzing censored data.

Chapter 9

Monte Carlo Simulation and Risk Assessment

9.1 Introduction

In the first eight chapters of this book we have discussed several statistical tools for looking at data, modeling probability distributions, estimating distribution parameters and quantiles, constructing prediction and tolerance intervals, comparing two or more groups, and testing for trend. All of our examples have concentrated on assessing how much chemical is in the environment and comparing chemical concentrations to "background." But given that chemicals are in the environment, what happens when people or other living organisms are exposed to these chemicals?

Of course, "chemicals" in the environment are part of our everyday lives: they are in the food we eat, the water we drink, the air we breathe. Some are natural and others are synthetic, having been added either on purpose or as a by-product of a manufacturing process. There is no doubt that the chemical revolution of twentieth century has improved our lives immensely. But we have also learned that some chemicals that improved some facet of our lives can have devastating consequences on our health and environment.

Several government agencies are charged with evaluating the potential health and ecological effects of environmental toxicants. Based on their assessments, these agencies set standards for acceptable concentration levels of these toxicants in air, water, soil, food, etc. The process of modeling exposure to a toxicant and predicting health or ecological effects is termed *risk assessment*, and *probabilistic risk assessment* uses probability distributions to characterize variability or uncertainty in risk estimates.

Risk assessment is a process where science, politics, and psychology all intersect. Not surprisingly, it is also a field full of controversy. References discussing risk assessment and the concept of risk include Byrd and Cothern (2000), Everitt (2008), Hallenbeck (1993), Laudan (1997), Lewis (1990), Lundgren and McMakin (2009), Neely (1994), Ostrom and Wilhelmsen (2012), Robson and Toscano (2007), Rodricks (2007), Suter et al. (2000), Suter (2007), USEPA (1992g, 1997a, b, c, 1999, 2001, 2005), Vose (2008), and Walsh (1996). In addition, the US Environmental Protection Agency has a web site dedicated to the topic of environmental risk assessment: www.epa.gov/risk. In this chapter, we will introduce basic mathematical models used in risk assessment and talk about how to use ENVSTATS and R to perform Monte Carlo simulation and probabilistic risk assessment (see Millard et al. 2014, for a more in-depth discussion of these topics).

9.2   Overview

Human and ecological risk assessment involves characterizing the exposure to a toxicant for one or several populations, quantifying the relationship between exposure (dose) and health or ecological effects (response), determining the risk (probability) of a health or ecological effect given the observed level(s) of exposure, and characterizing the uncertainty associated with the estimated risk (Hallenbeck 1993). Risk assessment is an enormous field of research and practice. The merits and disadvantages of particular exposure models (including fate and transport models) and dose-response models are not discussed in this chapter. Several journals, textbooks, and web sites are dedicated to risk assessment (see Introduction).

In the past, estimates of risk were often based solely on setting values of the input variables (e.g., body weight, dose, etc.) to particular point estimates and producing a single point estimate of risk, with little, if any, quantification of the uncertainty associated with the estimated risk. More recently, several practitioners have advocated "probabilistic" risk assessment, in which the input variables are considered random variables, so the result of the risk assessment is a probability distribution for predicted risk or exposure.

Usually, the equation describing risk or exposure is so complicated that it is not feasible to determine the output distribution using analytical methods, so the distribution of risk or exposure is derived via Monte Carlo simulation. This chapter discusses the concepts of Monte Carlo simulation, sensitivity and uncertainty analysis, and risk assessment, and shows you how to use ENVSTATS and R to perform probabilistic risk assessment.

9.3   Monte Carlo Simulation

Monte Carlo simulation is a method of investigating the distribution of a random variable by simulating random numbers (Gentle 1985). Usually, the random variable of interest, say Y, is some function of one or more other random variables:

$$Y = h(\underline{X}) = h(X_1, X_2, \ldots, X_k) \tag{9.1}$$

For example, Y may be an estimate of the median of a population with a Cauchy distribution, in which case the vector of random variables $\underline{X}$ represents k independent and identically distributed observations from some particular Cauchy distribution. As another example, Y may be the incremental lifetime cancer risk due to ingestion of soil contaminated with benzene (Thompson et al. 1992; Hamed and Bedient 1997). In this case the random vector $\underline{X}$ may represent observations from several kinds of distributions that characterize exposure and dose-response, such as benzene concentration in the soil, soil ingestion rate, average body weight, the cancer potency factor for benzene, etc. These distributions may or may not be assumed to be independent of one another (Smith et al. 1992; Bukowski et al. 1995).

Sometimes the input variables $X_1, X_2, \ldots, X_k$ are called *input parameters*. This terminology can be confusing, however, since the input variables are often random variables and therefore have ***distribution parameters*** associated with their probability distributions (e.g., mean and standard deviation for a normally distributed input variable).

Sometimes the distribution of Y in Eq. 9.1 can be derived analytically based on statistical theory (Springer 1979; Slob 1994). Often, however, the function h is complicated and/or the elements of the random vector $\underline{X}$ involve several kinds of probability distributions, making it difficult or impossible to derive the exact distribution of Y. In this case, Monte Carlo simulation can be used to approximate the distribution of Y. Monte Carlo simulation is often used in risk assessment, specifically in sensitivity and uncertainty analysis.

Monte Carlo simulation involves creating a large number of realizations of the random vector $\underline{X}$, say n, and computing Y for each of the n realizations of $\underline{X}$. The resulting distribution of Y, or some characteristic of this distribution (e.g., the mean), is then assumed to be "close" to the true distribution or distribution characteristic of Y. The adequacy of the approximation depends on a number of factors, including how well the mathematical relationship described in Eq. 9.1 reflects the true relationship between Y and $\underline{X}$, how well the specified distribution of $\underline{X}$ reflects its true distribution (including any possible dependencies between the individual elements of $\underline{X}$), and how many Monte Carlo samples or trials (n) are created. Usually, Monte Carlo simulation involves generating random numbers from some specified theoretical probability distribution, such as a normal, lognormal, beta, etc. When the simulation is done based on an empirical distribution, this is also called bootstrapping (Efron and Tibshirani 1993).

Various sources indicate that the term "Monte Carlo" comes from the code name of a World War II era project at Los Alamos Laboratories, although they differ on exactly who coined the term (Anderson 1986; Gentle 1985; Hayes 1993; Rubinstein 1981; Rugen and Callahan 1996). The code name comes from the casino in Monaco with the same name. References that address the issues of how to properly perform and report the results of a Monte Carlo simulation study include Burmaster and Anderson (1994), Hoaglin and Andrews (1975), Law (2006), and Vose (2008).

9.3.1 *Simulating the Distribution of the Sum of Two Normal Random Variables*

Suppose X_1 and X_2 are two independent standard normal random variables. Then the distribution of

$$Y = h(X_1, X_2) = X_1 + X_2 \qquad (9.2)$$

is normal with a mean of 0 and a variance of 2. Suppose, however, that we do not know how to derive the distribution of Y. We can use Monte Carlo simulation to investigate the shape of the distribution of Y, as well as compute characteristics of the distribution (e.g., mean, median, standard deviation, quantiles, etc.)

Figure 9.1 displays the empirical and true distribution of Y, where the empirical distribution of Y was derived by using Monte Carlo simulation with 100 trials.

That is, for each trial, two random numbers from a standard normal distribution were generated and added together. Figure 9.2 displays the empirical distribution based on 10,000 trials along with the true distribution. Table 9.1 displays some summary statistics for the two empirical distributions and compares them with the true population values. As we increase the number of Monte Carlo trials, the simulated distribution tends to get "closer" to the true distribution. This is called the *Law of Large Numbers*.

Parameter	Empirical (100)	Empirical (10,000)	Population N(0, 2)
Mean	0.12	0.03	0
Standard deviation	1.46	1.43	1.41
5th percentile	-2.10	-2.32	-2.33
95th percentile	2.97	2.41	2.33

Table 9.1 Comparison of empirical and population summary statistics

To generate the empirical distribution of the sum of two independent standard normal random variables based on 100 and 10,000 Monte Carlo trials, type these commands:

```
> df.100 <-data.frame(simulateMvMatrix(n = 100, seed = 20))

> y.100 <- with(df.100, Var.1 + Var.2)

> df.10000 <- data.frame(
    simulateMvMatrix(n = 10000, seed = 20)

> df.10000 <- with(df.10000, Var.1 + Var.2)
```

(You can also use the function mvrnom in the MASS package to create multivariate normal random numbers.) To create Fig. 9.1, type these commands:

```
> hist(y.100, freq = FALSE, col = "cyan",
    xlab = expression(paste("Y = ", X[1], " + ", X[2])),
    ylab = "Relative Frequency", main = "")

> pdfPlot(param.list = list(mean = 0, sd = sqrt(2)),
    add = TRUE, pdf.lwd = 3)
```

To create Fig. 9.2 type these commands:

```
> hist(y.10000, freq = FALSE, breaks = 75, col = "cyan",
    xlab = expression(paste("Y = ", X[1], " + ", X[2])),
    ylab = "Relative Frequency", main = "")

> pdfPlot(param.list = list(mean = 0, sd = sqrt(2)),
    add = TRUE, pdf.lwd = 3)
```

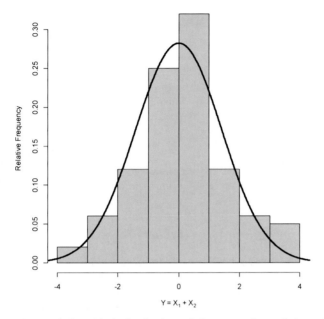

Fig. 9.1 Empirical and theoretical distribution of the sum of two independent N(0,1) random variables based on 100 Monte Carlo trials

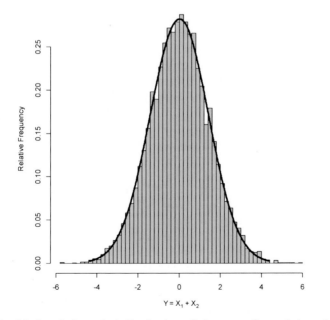

Fig. 9.2 Empirical and theoretical distribution of the sum of two independent N(0,1) random variables based on 10,000 Monte Carlo trials

9.4   Generating Random Numbers

A random number is a realization of a random variable, say X. For many people, the term random number initially conjures up an image of somehow choosing an integer between a specified lower and upper bound (e.g., 1 and 10), where each number is equally likely to be chosen. In that case, the random variable X is a discrete uniform random variable with probability density (mass) function given by:

$$f(x) = \Pr(X = x) = \frac{1}{10} \; ; \; x = 1, 2, \ldots, 10 \tag{9.3}$$

In general, a random number can be a realization of a random variable from any kind of probability distribution (e.g., uniform, normal, lognormal, gamma, empirical, etc.)

9.4.1   Generating Random Numbers from a Uniform Distribution

The R function `runif` generates **pseudo-random numbers** from a (continuous) uniform distribution. Random number generation in R is documented in the help file `Random`. The default generator is a Mersenne-Twister (Matsumoto and Nishimura 1998). References that discuss generating pseudo-random numbers include Barry (1996), Hayes (1993), Kennedy and Gentle (1980), Law (2006), Ripley (1981), and Rubinstein (1981).

Pseudo-random number generators start with an initial **seed**, and then appear to generate random numbers, although these numbers are actually generated by a deterministic mechanism. Each time you generate a set of random numbers, the value of the seed changes. If you start with the same seed, you will get the same sequence of pseudo-random numbers. You can use the R function `set.seed` to set the seed of the random number generator.

The **period** of a random number generator is the number of random numbers that can be generated before the sequence repeats itself. The period of the default generator in R is $2^{19,937} - 1$ (about $10^{6,000}$).

9.4.2   Generating Random Numbers from an Arbitrary Distribution

As we saw in Chap. 4, the R and ENVSTATS functions of the form `rabb` (where `abb` denotes the abbreviation of the distribution) generate random numbers from several theoretical probability distributions. For example, the function `rnorm` generates random numbers from a normal distribution.

To generate random numbers for a specified probability distribution, most computer software programs use the inverse transformation method (Law 2006; Rubinstein 1981; Vose 2008). Suppose the random variable U has a U[0,1] distribution, that is, a uniform distribution over the interval [0,1]. Let F_X denote the cumulative distribution function (cdf) of the specified probability distribution. Then the random variable X defined by:

$$X = F_X^{-1}(U) \tag{9.4}$$

has the specified distribution, where the quantity F_X^{-1} denotes the inverse of the cdf function F_X. Thus, to generate a set of random numbers from any distribution, all you need is a set of random numbers from a U[0,1] distribution and a function that computes the inverse of the cdf function for the specified distribution. Figure 9.3 illustrates the inverse transformation method for a standard normal distribution, with $U = 0.8$. In this case, the random number generated is $\Phi^{-1}(0.8)$, which is 0.8416212.

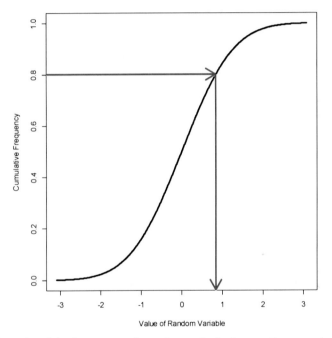

Fig. 9.3 Example of the inverse transformation method of generating a random number from the standard normal distribution

9.4.3 Latin Hypercube Sampling

Latin Hypercube sampling, sometimes abbreviated LHS, is a method of sampling from a probability distribution (one random variable) or a joint probability distribution (several random variables) that ensures all portions of the probability distribution are represented in the sample. It was introduced in the published literature by McKay et al. (1979). Other references include Iman and Conover (1980, 1982), and Vose (2008). Latin Hypercube sampling is an extension of quota sampling, and when applied to the joint distribution of k random variables, can be viewed as a k-dimensional extension of Latin square sampling, thus the name (McKay et al. 1979).

Latin Hypercube sampling was introduced to overcome the following problem in Monte Carlo simulation based on simple random sampling (SRS). Suppose we want to generate random numbers from a specified distribution. If we use simple random sampling, there is a low probability of getting very many observations in an area of low probability of the distribution. For example, if we generate n observations from the distribution, the probability that none of these observations falls into the upper 98th percentile of the distribution is 0.98^n. So, for example, there is a 13 % chance that out of 100 random numbers none will fall at or above the 98th percentile. If we are interested in reproducing the shape of the distribution, we will need a very large number of observations to ensure that we can adequately characterize the tails of the distribution (Vose 2008).

Latin Hypercube sampling was developed in the context of using computer models that required enormous amounts of time to run and for which only a limited number of Monte Carlo simulations could be implemented. In cases where it is fairly easy to generate tens of thousands of Monte Carlo trials, Latin Hypercube sampling may or may not offer any real advantage.

Latin Hypercube sampling works as follows for a single probability distribution. If we want to generate n random numbers from the distribution, the distribution is divided into n intervals of equal probability $1/n$. A random number is then generated from each of these intervals. For k independent probability distributions, LHS is applied to each distribution, and the resulting random numbers are matched at random to produce n random vectors of dimension k.

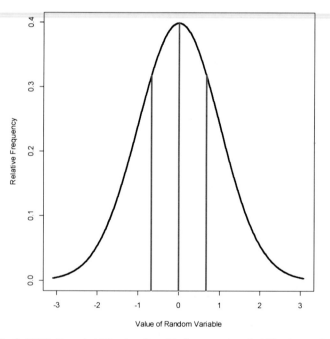

Fig. 9.4 N(0, 1) probability density with four equal-probability intervals

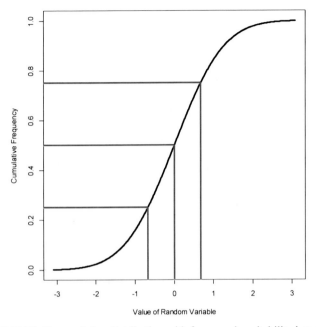

Fig. 9.5 N(0, 1) cumulative distribution with four equal-probability intervals

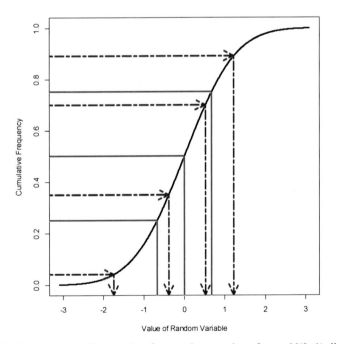

Fig. 9.6 Visual explanation of generating four random numbers from a N(0, 1) distribution using Latin Hypercube sampling

Figures 9.4, 9.5, and 9.6 illustrate Latin Hypercube sampling for a sample size of $n = 4$, assuming a standard normal distribution. Figure 9.4 shows the four equal-probable intervals for a standard normal distribution in terms of the probability density function, and Fig. 9.5 shows the same thing in terms of the cumulative distribution function. Figure 9.6 shows how Latin Hypercube sampling is accomplished using the inverse transformation method for generating random numbers. In this case, the interval [0,1] is divided into the four intervals [0, 0.25], [0.25, 0.5], [0.5, 0.75], and [0.75, 1]. Next, a uniform random number is generated within each of these intervals. For this example, the four numbers generated are (to two decimal places) 0.04, 0.35, 0.70, and 0.89. Finally, the standard normal random numbers associated with the inverse cumulative distribution function of the four uniform random numbers are computed: $-1.75, -0.39, 0.52$ and 1.23.

9.4.4   Example of Simple Random Sampling versus Latin Hypercube Sampling

Figure 9.7 displays a histogram of 50 observations based on a simple random sample from a standard normal distribution. Figure 9.8 displays the same thing based on a Latin Hypercube sample. You can see that the form of the histogram constructed with the Latin Hypercube sample more closely resembles the true underlying distribution.

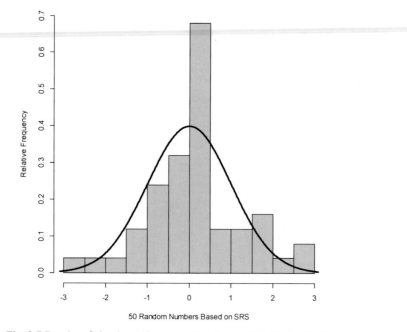

Fig. 9.7 Results of simple random sampling from a N(0, 1) distribution

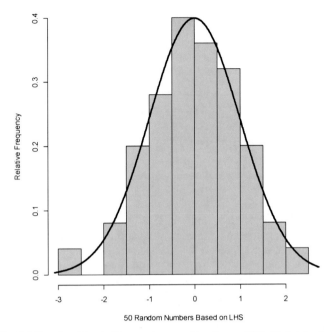

Fig. 9.8 Results of Latin Hypercube sampling from a N(0, 1) distribution

The EnvStats function `simulateVector` lets you generate random numbers from any of the built-in probability distributions in R and EnvStats using either simple random sampling or Latin hypercube sampling. To create Figs. 9.7 and 9.8 type these commands:

```
> x.srs <- simulateVector(50, seed = 798)

> hist(x.srs, freq = FALSE, breaks = 15, ylim = c(0, 0.7),
    col = "cyan", xlab = "50 Random Numbers Based on SRS",
    ylab = "Relative Frequency", main = "")

> pdfPlot(add = TRUE, pdf.lwd = 3)

> x.lhs <- simulateVector(50, seed = 798,
    sample.method = "LHS")

> hist(x.lhs, freq = FALSE, breaks = 15, ylim = c(0, 0.4),
    col = "cyan", xlab = "50 Random Numbers Based on LHS",
    ylab = "Relative Frequency", main = "")

> pdfPlot(add = TRUE, pdf.lwd = 3)
```

9.4.5   Properties of Latin Hypercube Sampling

Let Y denote the outcome variable for one trial of a Monte Carlo simulation, and suppose Y is a function of k independent random variables as shown in Eq. 9.1 above. McKay et al. (1979) consider the class of estimators of the form

$$T = T(\underline{Y}) = T(Y_1, Y_2, \ldots, Y_n) = \frac{1}{n} \sum_{i=1}^{n} g(Y_i) \qquad (9.5)$$

where g is an arbitrary function. This class of estimators includes the mean, the rth sample moment, and the empirical cumulative distribution function. Setting

$$\tau = E\left[T(\underline{Y})\right] \qquad (9.6)$$

McKay et al. (1979) show that under LHS, T is an unbiased estimator of τ, and also, if h in Eq. 9.1 is monotonic in each of its arguments and g is monotonic, then the variance of T under LHS is less than or equal to the variance of T under SRS.

Stein (1987) shows that the variance of the sample mean of Y under LHS is asymptotically less than the variance of the sample mean under simple random sampling whether or not the function h is monotonic in its arguments. Unfortunately, for most cases of LHS, the formula for the true variance of the sample mean is difficult to derive, and thus a good estimate of true variance is not available. Using the usual formula of dividing the sample variance by the sample size will usually overestimate the true variance of the sample mean.

Iman and Conover (1980) and Stein (1987) suggest producing several independent Latin Hypercube samples, say N, and for each Latin Hypercube sample computing the sample mean based on the n observations within that sample; they call this method *replicated Latin Hypercube sampling*. You can then estimate the variance of the sample mean by computing the usual sample variance of these N sample means. Note that this method can be applied to any quantity of interest, such as the median, 95th percentile, etc.

9.4.6   Generating Correlated Multivariate Random Numbers

Often, the input variables in a Monte Carlo simulation are known to be correlated, such as body weight and dermal area. If all of the input variables are normally distributed or all of them are lognormally distributed, you can easily generate correlated random numbers using the function `mvrnom` in the R package MASS. However, if the different input variables have different kinds of distributions, it is not straightforward how to generate correlated random variables. The EnvStats function `simulateMvMatrix` uses the method of Iman and Conover (1982) to allow you to generate a random sample or Latin Hypercube sample of correlated random variables from multiple types of distributions using rank correlations.

USEPA (2001, p. B-27) presents an example of a Monte Carlo simulation with correlated input variables for estimating the distribution in a certain human population of daily intake over a 30-year period of a chemical found in fish. The equation used to model chemical intake is given by:

$$Intake = \frac{CF \times IR \times FI \times EF \times ED}{BW \times AT} \tag{9.7}$$

and Table 9.2 displays the descriptions and assumed values or distributions for the input variables. The parameters for the lognormal distributions refer to the mean and standard deviation for the *untransformed* variable. The averaging time (AT) is simply the exposure duration multiplied by 365 days/year.

Variable	Description	Units	Point estimate or PDF
CF	Concentration in fish	µg/kg	25
IR	Fish ingestion rate	kg/meal	Lognormal (0.16, 0.07)
FI	Fraction ingestion from source	unitless	1.0
EF	Exposure frequency	meals/year	Lognormal (35.5, 25.0)
ED	Exposure duration	years	30
BW	Body weight	kg	70
AT	Averaging time	days	10,950

Table 9.2 Input variables, point estimates, and distributions for Eq. 9.7

To reproduce this example, first we will create a function for intake:

```
> Intake.fcn <- function(CF = 25, IR, FI = 1, EF, ED = 30,
    BW = 70, AT = ED * 365)
    { (CF * IR * FI * EF * ED) / (BW * AT) }
```

Next we will perform 5,000 Monte Carlo simulations of intake using Latin Hypercube sampling for each of four different scenarios of rank correlation between the input variables *IR* and *EF*: 0, 0.1, 0.5, and 0.9.

```
> cors <- c(0, 0.1, 0.5, 0.9)

> Intake.mat <- matrix(as.numeric(NA), nrow = 5000, ncol = 4,
    dimnames = list(NULL, paste("Cor", cors, sep = ".")))

> for(j in 1:4) {
    IR.EF.df <- data.frame(simulateMvMatrix(5000,
    distributions = c(IR = "lnormAlt", EF = "lnormAlt"),
    param.list = list(IR = list(mean = 0.16, cv = 0.07/0.16),
    EF = list(mean = 35.5, cv = 25/35.5)),
    cor.mat = matrix(c(1, cors[j], cors[j], 1), ncol = 2),
    sample.method = "LHS", seed = 428))

    Intake.mat[, j] <- with(IR.EF.df,
        Intake.fcn(IR = IR, EF = EF))
    }
```

Finally, here are the summary statistics showing the mean and the 50th, 95th and 97.5th percentiles for each scenario. Note that intake is in units of $\mu g/(kg\text{-day})$, however USEPA (2001) reports the results of the simulation in units of $\mu g/day$, so we will multiply our results by 70 kg (the assumed body weight).

```
> Results.mat <- 70 * apply(Intake.mat, 2,
    function(x) c(Mean = mean(x),
    quantile(x, probs = c(0.5, 0.95, 0.975)))))

> round(Results.mat, 2)
```

	Cor.0	Cor.0.1	Cor.0.5	Cor.0.9
Mean	0.39	0.40	0.45	0.49
50%	0.29	0.29	0.29	0.29
95%	1.04	1.10	1.32	1.57
97.5%	1.33	1.42	1.77	2.17

Note that these results differ from the results presented in USEPA (2001) by about a factor of 1/4. As USEPA (2001) points out, positive rank correlations have little effect on the median of the distribution for intake but tend to widen the tails of the distribution.

9.5   Uncertainty and Sensitivity Analysis

Uncertainty analysis and sensitivity analysis are terms used to describe various methods of characterizing the behavior of a complex mathematical/computer model. The model in Eq. 9.1 above is different from most conventional statistical models, where the form of the model is:

$$Y = h(\underline{X}) + \varepsilon \qquad (9.8)$$

(e.g., linear regression models, generalized linear models, nonlinear regression models, etc.). In Eq. 9.8, the vector $\underline{X}$ is assumed to be set or observed at fixed values, and for fixed values of $\underline{X}$ the response variable Y deviates about its mean value according to the distribution of the error term ε. This kind of model is useful when we are interested in the specific relationship between Y and $\underline{X}$, and we want to predict the value of Y for a specified value of $\underline{X}$. Furthermore, this kind of model is fit using paired observations of Y and $\underline{X}$.

In Eq. 9.1, Y is assumed to be observed without error, that is, the value of Y is deterministic for a set value of $\underline{X}$. The output variable Y, however, is a random variable when some or all of the input variables X_1, X_2, ..., X_k are random variables. This kind of model is useful when we are interested in describing the distribution of Y taken over all possible (read as "reasonable and realistic") combinations of the input variables. Furthermore, paired observations of Y and $\underline{X}$ are usually not available to validate this kind of model, hence, there is some amount of unquantifiable uncertainty associated with this kind of model.

Uncertainty analysis involves describing the variability or distribution of values of the output variable Y that is due to the collective variation in the input variables $\underline{X}$ (Iman and Helton 1988). This description usually involves graphical displays such as histograms, empirical density plots, and empirical cdf plots, as well as summary statistics such as the mean, median, standard deviation, coefficient of variation, 95th percentile, etc.

Sensitivity analysis involves determining how the distribution of Y changes with changes in the individual input variables. It is used to identify which input variables contribute the most to the variation or uncertainty in the output variable Y (Iman and Helton 1988). Sensitivity analysis is also used in a broader sense to determine how changing the distributions of the input variables and/or their assumed correlations or even changing the form of the model affects the output (Thompson et al. 1992; Smith et al. 1992; Cullen 1994; Shlyakhter 1994; Bukowski et al. 1995; Hamed and Bedient 1997; USEPA 1997a).

9.5.1   *Important Versus Sensitive Parameters*

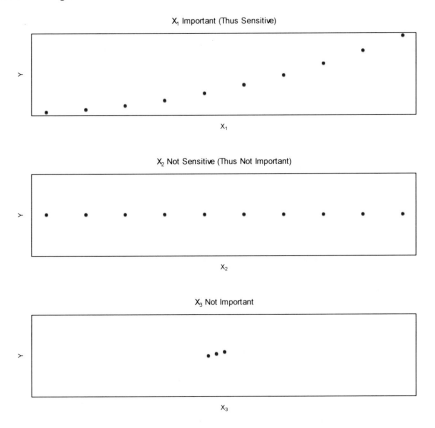

Fig. 9.9 Three examples of the concepts of "important" and "sensitive" parameters

Two useful concepts associated with sensitivity analysis are important parameters (variables) and sensitive parameters (variables) (Crick et al. 1987 as cited in Hamby 1994). **Sensitive parameters** have a substantial influence on the resulting distribution of the output variable Y, that is, small changes in the value of a sensitive parameter result in substantial changes in Y. **Important parameters** have some amount of uncertainty and/or variability associated with them and this variability contributes substantially to the resulting variability in the output variable Y.

Figure 9.9 illustrate these two concepts for the simple case of three input variables. The top plot is an example of an important variable. An important variable is always sensitive. The middle plot is an example of a variable that is not sensitive, and hence not important. The bottom plot is an example of a variable that is not important. This variable may not be sensitive like the one in the middle plot, or it may be sensitive like the one in the top plot but it is not important because of its limited variability.

9.5.2 Uncertainty Versus Variability

The terms *uncertainty* and *variability* have specific meanings in the risk assessment literature that do not necessarily match their meanings in the statistical literature or everyday language. The term **variability** refers to the inherent heterogeneity of a particular variable (parameter). For example, there is natural variation in body weight and height between individuals in a given population. The term **uncertainty** refers to a lack of knowledge about specific parameters, models, or factors (Morgan and Henrion 1990; Hattis and Burmaster 1994; Rowe 1994; Bogen 1995; USEPA 1997a, 2001). Uncertainty can be classified into three broad categories:

- **Parameter uncertainty**. Uncertainty in the point estimates or distribution parameters used to estimate variables of the model. For example, we may be uncertain about the true distribution of exposure to a toxic chemical in a population (parameter uncertainty due to lack of data, measurement errors, sampling errors, systematic errors, etc.).
- **Model uncertainty**. Uncertainty in the adequacy of the model. We may be uncertain how well our model of incremental lifetime cancer risk reflects reality (model uncertainty due to simplification of the process, misspecification of the model structure, model misuse, use of inappropriate surrogate variables, etc.),
- **Scenario uncertainty**. Uncertainty regarding missing or incomplete information to fully define what we are modeling (e.g., exposure). For example, we may be uncertain about whether a chemical is even present at a site of concern (scenario uncertainty due to descriptive errors, aggregation errors, errors in professional judgment, incomplete analysis, etc.).

We can usually reduce uncertainty through further measurement or study. We cannot reduce variability, since it is inherent in the variable. Note that in the risk assessment literature, measurement error contributes to uncertainty; we can decrease uncertainty by decreasing measurement error. In the statistical literature, measurement error is one component of the variance of a random variable. Note that a parameter (input variable) may have little or no variability associated with it, yet still have uncertainty associated with it (e.g., the speed of light is constant, but we only know its value to a given number of decimal places).

The terms **uncertainty** and **uncertainty analysis** should not be confused. Uncertainty analysis characterizes the distribution of the output variable Y. The output variable Y varies due to the fact that the input variables are random variables. The distributions of the input random variables reflect both variability (inherent heterogeneity) and uncertainty (lack of knowledge).

9.5.3   Sensitivity Analysis Methods

Sensitivity analysis methods can be classified into three groups: one-at-a-time deviations from a baseline case, factorial design and response surface modeling, and Monte Carlo simulation (Hamby 1994). Each of these kinds of sensitivity analysis is briefly discussed below. For more detailed information, see Vose (2008), USEPA (2001), and Millard et al. (2014). Several studies indicate that using Monte Carlo simulation in conjunction with certain sensitivity measures usually provides the best method of determining sensitivity of the parameters.

One-at-a-Time Deviations from a Baseline Case

These sensitivity analysis methods include differential analysis and measures of change in output to change in input. Differential analysis is simply approximating the variance of the output variable Y at a particular value of the input vector $\underline{X}$ (called the **baseline case**) by using a first-order Taylor series expansion (Kotz and Johnson 1985, Volume 8; Downing et al. 1985; Seiler 1987; Iman and Helton 1988; Hamby 1994). This approximating equation for the variance of Y is useful for quantifying the proportion of variability in Y that is accounted for by each input variable. Unfortunately, the approximation is usually good only in a small region close to the baseline case, and the relative contribution of each input variable to the variance of Y may differ dramatically for differently chosen baseline cases. Also, differential analysis requires the calculation of partial derivatives, which may or may not be a simple task, depending on the complexity of the input function h in Eq. 9.1.

Measures of change in output to change in input include the ratio of percent change in Y to percent change in X_i (Hamby 1994), the ratio of percent change in Y to change in X_i in units of the standard deviation of X_i (Hamby 1994; Finley and Paustenbach 1994), the percent change in Y as X_i ranges from its minimum to maximum value (Hamby 1994), and spider plots, which are plots of Y versus percent change in X_i, or Y versus percentiles of X_i (Vose 2008).

Factorial Design and Response Surface Modeling

The concepts of factorial designs and response surfaces come from the field of experimental design (Box et al. 1978). In the context of sensitivity analysis for computer models, n distinct values of the input vector $\underline{X}$ are chosen (usually reflecting the possible ranges and medians of each of the k input variables), and the model output Y is recorded for each of these input values. Then a multiple linear regression model (a ***response surface***) is fit to these data. The fitted model is called the fitted response surface, and this response surface is used as a replacement for the computer model (Downing et al. 1985; Iman and Helton 1988; Hamby 1994). The sensitivity analysis is based on the fitted response surface. The reason for using a response surface to replace the actual model output is that some computer models are, or used to be, very costly to run, whereas computing output based on a response surface is relatively inexpensive.

One way to rank the importance of the variables in the response surface is to simply compare the magnitudes of the estimated coefficients. The estimated coefficients, however, depend on the units of the predictor variables in the model, so some sources suggest using standardized regression coefficients (Iman and Helton 1988; Hamby 1994). The standardized regression coefficients are simply the coefficients that are obtained from fitting the response surface model based on the "standardized" output variable and the "standardized" predictor (input) variables. That is, for each variable, each observation is replaced by subtracting the mean (for that variable) from the observation and dividing by the standard deviation (for that variable).

A big problem with using standardized coefficients to determine the importance of predictor variables is that they depend on the range of the predictor variables (Weisberg 1985). So, for example, the variable X_3 in Fig. 9.9 above may be very important but it has a very limited range and therefore does not contribute to very much variation in Y.

Iman and Helton (1988) compared uncertainty and sensitivity analysis of several models based on differential analysis, factorial design with a response surface model, and Monte Carlo simulation using Latin Hypercube sampling. The main outcome they looked at for uncertainty analysis was estimating the cumulative distribution function. They found that the models were too mathematically complex to be adequately represented by a response surface. Also, the results of differential analysis gave widely varying results depending on the values chosen for the baseline case. The method based on Monte Carlo simulation gave the best results.

Monte Carlo Simulation

Monte Carlo simulation is used to produce a distribution of Y values based on generating a large number of values of the input vector $\underline{X}$ according to the joint distribution of $\underline{X}$. There are several possible ways to produce a distribution for Y, including varying all of the input parameters (variables) simultaneously, varying one parameter at a time while keeping the others fixed at baseline values, or varying the parameters in one group while keeping the parameters in the other groups at fixed baseline values. Sensitivity methods that can be used with Monte Carlo simulation results include the following:

- **Histograms, Empirical CDF Plots, Percentiles of Output**. A simple graphical way to assess the effect of different input variables or groups of input variables on the distribution of Y is to look at how the histogram and empirical cdf of Y change as you vary one parameter at a time or vary parameters by groups (Thompson et al. 1992). Various quantities such as the mean, median, 95th percentile, etc. can be displayed on these plots as well.

- **Scatterplots**. Another simple graphical way to assess the effect of different input variables on the distribution of Y and their relationship to each other is to look at pair-wise scatterplots.

- **Correlations and Partial Correlations**. A quantitative measure of the relationship between Y and an individual input variable X_i is the correlation between these two variables, computed based on varying all parameters simultaneously (Saltelli and Marivoet 1990; Hamby 1994). Vose (2008) suggests using *tornado charts*, which are simply horizontal barcharts displaying the values of the correlations. Individual correlations are hard to interpret when some or most of the input variables are highly related to one another. One way to get around this problem is to look at partial correlation coefficients. See Millard et al. (2014) for more information.

- **Change in Output to Change in Input**. Any of the types of measures that are described above under the section *One-at-a-Time Deviations from a Baseline Case* can be adapted to the results of a Monte Carlo simulation. Additional measures include relative deviation, in which you vary one parameter at a time and compute the coefficient of variation (CV) of Y for each case, and relative deviation ratio, in which you vary one parameter at a time and compute the ratio of the CV of Y to the CV of X_i (Hamby 1994).

- **Response Surface**. This methodology that was described above in the section *Factorial Design and Response Surface Modeling* can be adapted to the results of a Monte Carlo simulation. In this case, use the model input and output to fit a regression equation (possibly stepwise) and then use standardized coefficients to rank the input variables (Iman and Helton 1988; Saltelli and Marivoet 1990).

- **Comparing Groupings within Input Distributions Based on Partitioning the Output Distribution**. One final method of sensitivity analysis that has been used with Monte Carlo simulation is to divide the distribution of the output variable Y into two or more groups, and then to compare the distributions of an input variable that has been split up based on these groupings (Saltelli and Marivoet 1990; Hamby 1994; Vose 2008). For example, you could divide the distribution of an input variable X_i into two groups based on whether the values yielded a value of Y below the median of Y or above the median of Y. You could then compare the distributions of these two groups, compare these distributions with a goodness-of-fit test, or compare the means or medians of these distributions with the t-test or Wilcoxon rank sum test. A significant difference between the two distributions is an indication that the input variable is important in determining the distribution of Y.

9.5.4   Uncertainty Analysis Methods

A specific model such as Eq. 9.1 with specific joint distributions of the input variables leads to a specific distribution of the output variable. The process of describing the distribution of the output variable is called **uncertainty analysis** (Iman and Helton 1988). This description usually involves graphical displays such as histograms, empirical distribution plots, and empirical cdf plots, as well as summary statistics such as the mean, median, standard deviation, coefficient of variation, 95th percentile, etc.

Sometimes the distribution of Y in Eq. 9.1 can be derived analytically based on statistical theory (Springer 1979; Slob 1994). For example, if the function h describes a combination of products and ratios, and all of the input variables have a lognormal distribution, then the output variable Y has a lognormal distribution as well, since products and ratios of lognormal random variables have lognormal distributions. Many risk models, however, include several kinds of distributions for the input variables, and some risk models are not easily described in a closed algebraic form. In these cases, the exact distribution of Y can be difficult or almost impossible to derive analytically.

The rest of this section briefly describes some methods of uncertainty analysis based on Monte Carlo simulation. For more information on uncertainty analysis, see Vose (2008), USEPA (2001), and Millard et al. (2014).

Quantifying Uncertainty with Monte Carlo Simulation

When the distribution of Y cannot be derived analytically, it can usually be estimated via Monte Carlo simulation. Given this simulated distribution, you can construct histograms or empirical density plots and empirical cdf plots, as well as compute summary statistics. You can also compute confidence bounds for specific quantities, such as percentiles. These confidence bounds are based on the assumption that the observed values of Y are randomly selected based on simple random sampling. When Latin Hypercube sampling is used to generate input variables and hence the output variable Y, the statistical theory for confidence bounds based on simple random sampling is not truly applicable (Easterling 1986; Iman and Helton 1991; Stein 1987). Most of the time, confidence bounds that assume simple random sampling but are applied to the results of Latin Hypercube sampling will probably be too wide.

Quantifying Uncertainty by Repeating the Monte Carlo Simulation

One way around the above problem with Latin Hypercube sampling is to use replicated Latin Hypercube sampling, that is, repeat the Monte Carlo simulation numerous times, say N, so that you have a collection of N empirical distributions of Y, where each empirical distribution is based on n observations of the input vector $\underline{X}$. You can then use these N replicate distributions to assess the variability of the sample mean, median, 95th percentile, empirical cdf, etc.

A simpler process is to repeat the simulation just twice and compare the values of certain distribution characteristics, such as the mean, median, 5th, 10th, 90th and 95th percentiles, and also to graphically compare the two empirical cdf plots. If the values between the two simulations are within a small percentage of each other, then you can be fairly confident about characterizing the distribution of the output variable. Iman and Helton (1991) did this for a very complex risk assessment for a nuclear power plant and found a remarkable agreement in the empirical cdf's. Thompson et al. (1992) did the same thing for a risk assessment for incremental lifetime cancer risk due to ingestion or dermal contact with soil contaminated with benzene.

You may also want to compare the results of the original simulation with a simulation that uses say twice as many Monte Carlo trials (e.g., Thompson et al., 1992). Barry (1996) warns that if the moments of the simulated distribution do not appear to stabilize with an increasing number of Monte Carlo trials, this can mean that they do not exist. For most risk models, however, the true distributions of any random variables involved in a denominator in Eq. 9.1 are bounded above 0, so the moments will exist. If a random variable involved in the denominator has a mean or median that is close to 0, it is important to use a bounded or truncated distribution to assure the random variable stays sufficiently far away from 0.

Quantifying Uncertainty Based on Modeling Distribution Parameter Uncertainty

To account for the uncertainty in specifying the distribution of the input variables, some authors suggest using mixture distributions to describe the distributions of the input variables (e.g., Hoffman and Hammonds 1994; Burmaster and Wilson 1996; Vose 2008). For each input variable, a distribution is specified for the parameter(s) of the input variable's distribution. Some authors call random variables with this kind of distribution *second-order random variables* (e.g., Burmaster and Wilson 1996). Vose (2008) calls this *second-order fitting*. USEPA (2001) distinguishes between one-dimensional Monte Carlo analysis (1-D MCA) in which some or all of the input variables are characterized by probability distributions with fixed parameters versus two-dimensional Monte Carlo analysis (2-D MCA) in which one or more parameters for some of the probability distributions are not fixed but themselves random variables from some specified distribution.

For example we may assume the first input variable comes from a lognormal distribution with a certain mean and coefficient of variation (CV). The mean is unknown to a certain degree, and so we may specify that the mean comes from a uniform distribution with a given set of upper and lower bounds. We can also specify a distribution for the CV. In this case, the Monte Carlo simulation can be broken down into two stages. In the first stage, a set of parameters is generated for each input distribution. In the second stage, n realizations of the input vector $\underline{X}$ are generated based on this one set of distribution parameters. This two-stage process is repeated N times, so that you end up with N different empirical distributions of the output variable Y, and each empirical distribution is based on n observations of the input vector $\underline{X}$.

For a fixed number of Monte Carlo trials nN, the optimal combination of n and N will depend on how the distribution of the output variable Y changes relative to variability in the input distribution parameter(s) versus variability in the input variables themselves. For example, if the distribution of Y is very sensitive to changes in the input distribution parameters, then N should be large relative to n. On the other hand, if the distribution of Y is relatively insensitive to the values of the parameters of the input distributions, but varies substantially with the values of the input variables, then N may be small relative to n.

9.5.5 *Caveat*

An important point to remember is that no matter how complex the mathematical model in Eq. 9.1 is, or how extensive the uncertainty and sensitivity analyses are, there is always the question of how well the mathematical model reflects reality. The only way to attempt to answer this question is with data collected directly on the input and output variables. Often, however, it is not possible to do this, which is why the model was constructed in the first place.

For example, in order to attempt to directly verify a model for incremental lifetime cancer risk for a particular exposed population within a particular geographical region, you have to collect data on lifetime exposure for each person and the actual proportion of people who developed that particular cancer within their lifetime, accounting for competing risks as well, and compare these data to similar data collected on a proper control population. A controlled experiment that involves exposing a random subset of a particular human population to a toxin and following the exposed and control group throughout their lifetimes for the purpose of a risk assessment is not possible to perform for several reasons, including ethical and practical ones. Rodricks (2007) is an excellent text that discusses the complexities of risk assessment based on animal bioassay and epidemiological studies.

9.6   Risk Assessment

This section discusses the concepts and practices involved in risk assessment, and gives examples of how to use ENVSTATS to perform probabilistic risk assessment.

9.6.1 *Definitions*

It will be helpful to start by defining common terms and concepts used in risk assessment.

Risk

The common meaning of the term *risk* when used as a noun is "the chance of injury, damage, or loss." Thus, risk is a probability, since "chance" is another term for "probability."

Risk Assessment

Risk assessment is the practice of gathering and analyzing information in order to predict future risk. Risk assessment has been commonly used in the fields of insurance, engineering, and finance for quite some time. In the last couple of decades it has been increasingly applied to the problems of predicting human health and ecological effects from exposure to toxicants in the environment (e.g., Hallenbeck 1993; Suter 2007). In this chapter, the term risk assessment is applied in the context of human health and ecological risk assessment.

The basic model that is often used as the foundation for human health and ecological risk assessment is:

$$Risk = Dose \times \Pr(Effect \text{ per } Unit\,Dose)$$
$$= Intake \times CSF \tag{9.9}$$

That is, the risk of injury (the **effect**) to an individual is equal to the amount of toxicant the individual absorbs (the **dose** or **intake**) times the probability of the effect occurring for a single unit of the toxicant. If the effect is some form of cancer, the second term on the right-hand side of Eq. 9.9 is often called the **cancer slope factor** (abbreviated CSF) or the **cancer potency factor** (abbreviated CPF).

The first term on the right-hand side of Eq. 9.9, the dose, is estimated by identifying sources of the toxicant and quantifying their concentrations, identifying how these sources will expose an individual to the toxicant (via fate and transport models), quantifying the amount of exposure an individual will receive, and estimating how much toxicant the individual will absorb at various levels of exposure. Sometimes the dose represents the amount of toxicant absorbed over a lifetime, and sometimes it represents the amount absorbed over a shorter period of time.

The second term on the right-hand side of Eq. 9.9, the probability of an effect (CSF or CPF), is estimated from a **dose-response curve**, a model that relates the probability of the effect to the dose received. Dose-response curves are developed from controlled laboratory experiments on animals or other organisms, and/or from epidemiological studies of human populations (Hallenbeck 1993; Piegorsch and Bailer 2005).

Risk assessment involves four major steps (Hallenbeck 1993; USEPA 1995c, 2005; Piegorsch and Bailer 2005):

- **Hazard Identification**. Describe the effects (if any) of the toxicant on laboratory animals, humans, and/or wildlife species, based on documented studies. Describe the quality and relevance of the data from these studies. Describe what is known about how the toxicant produces these effects. Describe the uncertainties and subjective choices or assumptions associated with determining the degree of hazard of the toxicant.
- **Dose-Response Assessment**. Describe what is known about the biological mechanism that causes the health or ecological effect. Describe what data, models, and extrapolations have been used to develop the dose-response curve for laboratory animals, humans, and/or wildlife species.

Describe the routes and levels of exposure used in the studies to determine the dose-response curve, and compare them to the expected routes and levels of exposure in the population(s) of concern. Describe the uncertainties and subjective choices or assumptions associated with characterizing the dose-response relationship.

- **Exposure Assessment**. Identify the sources of environmental exposure to the population(s) of concern. Describe what is known about the principal paths, patterns, and magnitudes of exposure. Determine average and "high end" levels of exposure. Describe the characteristics of the populations that are potentially exposed. Determine how many members of the population are likely to be exposed. Describe the uncertainties and subjective choices or assumptions associated with characterizing the exposure for the population of concern.

- **Risk Characterization**. Incorporate all of the information from the hazard identification, dose-response assessment, and exposure assessment steps into a single assessment of the overall risk of the toxicant. Usually some form of Eq. 9.9 is used to estimate the risk for a particular population of concern. Both sensitivity analysis and uncertainty analysis should be applied to the risk assessment model to quantify the uncertainty associated with the estimated risk. USEPA (2005, p. 5-1) states: "The risk characterization includes a summary for the risk manager in a nontechnical discussion that minimizes the use of technical terms. It is an appraisal of the science that informs the risk manager in public health decisions, as do other decision-making analyses of economic, social, or technology issues."

Risk Assessment Versus Risk Characterization

USEPA (1995c) distinguishes between the process of risk assessment and risk characterization. *Risk characterization* is the summarizing step of risk assessment that integrates all of the information from the risk assessment, including uncertainty and sensitivity analyses and a discussion of uncertainty versus variability, to form an overall conclusion about the risk. Risk assessment is the tool that a risk assessor uses to produce a risk characterization. A risk characterization is the product that is delivered to the risk assessor's client: the risk manager.

Risk Management

Risk management is the process of using information from risk characterizations (calculated risks), perceived risks, regulatory policies and statutes, and economic and social analyses in order to make and justify a decision (USEPA 1995c). If the risk manager decides that the risk is not acceptable, he or she will order or recommend some sort of action to decrease the risk. If the risk manager decides that the risk poses minimal danger to the population of concern, he or she may recommend that no further action is needed at the present time.

Because the risk manager is the client of the risk assessor, the risk manager must be involved in the risk assessment process from the start, helping to determine the scope and endpoints of the risk assessment (USEPA 1995c). Also, the risk manager must interact with the risk assessor throughout the risk assessment process, so that he or she may take responsibility for critical decisions. The risk manager, however, must be careful not to let non-scientific (e.g., political) issues influence the risk assessment. Non-scientific issues are dealt with at the risk management stage, not the risk assessment stage.

Risk Communication

USEPA (1995c) defines **risk communication** as exchanging information with the public. While the communication of risk from the risk assessor to the risk manager is accomplished through risk characterization, the communication of risk between the risk manager (or representatives of his or her agency) and the public is accomplished through risk communication. The risk characterization will probably include highly technical information, while risk communication should concentrate on communicating basic ideas of risk to the public.

9.6.2   Building a Risk Assessment Model

Usually some form of Eq. 9.9 is used to estimate the risk for a particular population of concern. The form of the two terms on the right-hand side of Eq. 9.9 may be very complex. Estimation of dose involves identifying sources of exposure, postulating pathways of exposure from these sources, estimating exposure concentrations, and estimating the resulting dose for a given exposure. Estimation of dose-response involves using information from controlled laboratory experiments on animals and/or epidemiological studies. Given a set of dose-response data, there are several possible statistical models that can be used to fit these data, including tolerance distribution models, mechanistic models, linear-quadratic-exponential models, and time-to-response models (Hallenbeck 1993).

Probably the biggest controversy in risk assessment involves the extrapolation of dose-response data from high-dose to low-dose and from one species to another (e.g., between mice and humans). Rodricks (2007) discusses these problems in detail. An example of this problem is the case of saccharin, which was shown in the late 1970s to produce bladder tumors in male rats that were fed extremely large concentrations of the chemical. These studies led the FDA to call for a ban on saccharin, but Congress placed a moratorium on the ban that was renewed periodically. A little over two decades later, the National Institute of Environmental Health Sciences stated that new studies show "no clear association" between saccharin and human cancer (*The Seattle Times*, Tuesday, May 16, 2000) and took saccharin off of its list of cancer-causing chemicals.

Many risk assessment models have the general form:

$$Risk = Dose \times \Pr\left(Effect \ \text{per} \ Unit \ Dose\right)$$

$$= h\left(X_1, X_2, \ldots, X_k\right) \times CPF$$

(9.10)

That is, the risk is assumed to be proportional to dose, and the dose term is a function of several input variables. For example, USEPA (2001) uses the following equation for chronic daily intake:

$$Intake = \frac{C \times IR \times EF \times ED}{BW \times AT}$$

(9.11)

where C is the chemical concentration, IR is the ingestion or contact rate, EF is the exposure frequency, ED is the exposure duration, BW is body weight, and AT is the averaging time (equal to exposure duration $\times$ 365 days/year for non-carcinogens and 70 years $\times$ 365 days/year for carcinogens).

Usually, many of the input variables and sometimes the cancer potency factor (CPF) in Eq. 9.10 are themselves assumed to be random variables because they exhibit inherent heterogeneity (variability) within the population (e.g., body weight, fluid intake, etc.), and because there is a certain amount of uncertainty associated with their values. The choice of what distribution to use for each of the input variables is based on a combination of available data and expert judgment. USEPA (2001, pp. 1–13) states that a convenient aid to understanding the Monte Carlo approach to risk assessment is to think of each iteration as representing a single individual, and the collection of iterations as representing the population. Thus, "Each iteration of a Monte Carlo simulation should represent a plausible combination of input values (i.e., exposure and toxicity variables), which may require using bounded or truncated probability distributions …"

9.6.3   Example: Quantifying Variability and Parameter Uncertainty

USEPA (2001, pp. 3–13) presents an example of quantifying variability and parameter uncertainty in a probabilistic risk assessment involving exposure to a chemical via soil ingestion (obviously based on an example from Thompson et al., 1992). The risk equation is given by:

$$Risk = \frac{C \times IR \times CF \times EF \times ED}{BW \times AT} \times CSF_{oral}$$

(9.12)

and Table 9.3 displays the descriptions and assumed values or distributions for the input variables for four different cases. Two sources of variability are quantified: (1) inter-individual variability in exposure frequency (EF), characterized by a triangular distribution, and (2) inter-individual variability in exposure duration (ED), characterized by a truncated lognormal distribution. In addition, two sources of uncertainty are presented: (1) a point estimate for soil and dust ingestion rate (IR), intended to characterize the reasonable maximum exposure (RME), and (2) an upper truncation limit of the lognormal distribution for *ED*, intended to represent a plausible upper bound for the exposed population.

Variable	Description	1-D MCA			2-D MCA
		Case 1	Case 2	Case 3	Case 4
C	Concentration (mg/kg)	500	500	500	500
IR	Ingestion rate (mg/day)	50	100	200	Uniform min = 50 max = 200
CF	Conversion factor (kg/mg)	1e-6	1e-6	1e-6	1e-6
EF	Exposure frequency (days/year)	Triangular Min = 200 Mode = 250 Max = 350	Triangular Min = 200 Mode = 250 Max = 350	Triangular Min = 200 Mode = 250 Max = 350	Triangular min = 200 mode = 250 max = 350
ED	Exposure duration (years)	Truncated lognormal Mean = 9 cv = 10/9 Max = 26	Truncated lognormal Mean = 9 cv = 10/9 Max = 33	Truncated lognormal Mean = 9 cv = 10/9 Max = 40	Truncated lognormal Mean = 9 cv = 10/9 Max ~ uniform Min = 26 Max = 40
BW	Body weight (kg)	70	70	70	70
AT	Averaging time (days)	25,550	25,550	25,550	25,550
CSF	Cancer slope factor $(\text{mg/kg-day})^{-1}$	1e-1	1e-1	1e-1	1e-1

Table 9.3 Input variables, point estimates, and distributions for Eq. 9.12

The first three cases involve one-dimensional Monte Carlo Analysis (1-D MCA) in which the input variables *EF* and *ED* have associated triangular and truncated lognormal probability distributions, respectively, with fixed parameters within each case. The value of the point estimate for *IR* varies between Case 1, 2, and 3, as does the value of the parameter "max" in the truncated lognormal distribution for *ED*. The fourth case involves two-dimensional Monte Carlo analysis (2-D MCA) in which the parameter "max" in the truncated lognormal distribution for *ED* is itself allowed to vary according to a uniform distribution. Note also that the variable *IR* now has an associated probability distribution in this case

For Cases 1–3, simulations were run with 10,000 iterations and Latin Hypercube sampling. Figure 9.10 shows the results of these simulations by plotting the empirical cdf for each case. Each simulation used a different combination of plausible estimates of the reasonable maximum exposure (RME) value for *IR* and the upper truncation limit for *ED*, as discussed above. The results provide a bounding estimate on the risk distribution given these two sources of uncertainty. The 95th percentile risk, highlighted as an example of the RME risk estimate, may range from approximately 7.2×10^{-6} and 3.4×10^{-5}.

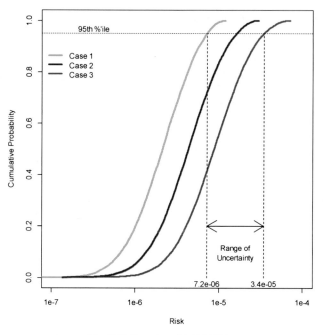

Fig. 9.10 Results of 1-D MCA

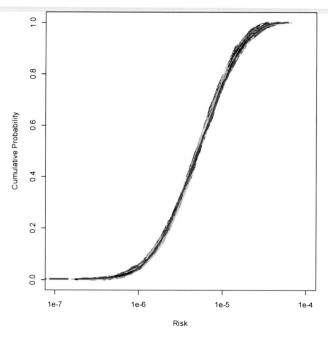

Fig. 9.11 Results of 2-D MCA

For Case 4, the 2-D MCA was performed by first generating 250 random values of the "max" parameter of the truncated lognormal distribution for *ED* using Latin Hypercube sampling (250 iterations of the outer loop representing uncertainty), and then for each of these 250 values, 2,000 iterations of the inner loop (representing variability) were run. Figure 9.11 shows the 250 empirical cdf plots; you can see that for this scenario the uncertainty in the "max" parameter has a very small effect on distribution of risk compared to the differences in the distribution of risk we saw in Cases 1–3.

To create Fig. 9.10, type these commands:

```
> Risk.fcn <- function(C = 500, IR, CF = 1e-6, EF, ED,
    BW = 70, AT = 25550, CSF = 0.1) {
    CSF * (C * IR * CF * EF * ED) / (BW * AT)
    }

> IR.vec <- c(50, 100, 200)

> ED.max <- c(26, 33, 40)

> Risk.mat <- matrix(as.numeric(NA), nrow = 10000, ncol = 3)

> set.seed(398)

> for(j in 1:3) {
    EF <- simulateVector(n = 10000, distribution = "tri",
    param.list = list(min = 200, mode = 250, max = 350),
    sample.method = "LHS")

    ED <- simulateVector(n = 10000, distribution =
    "lnormTruncAlt", param.list = list(mean = 9, cv = 10/9,
    min = 0, max = ED.max[j]), sample.method = "LHS")

    Risk.mat[, j] <- Risk.fcn(IR = IR.vec[j], EF = EF,
    ED = ED)
    }

> ecdfPlot(log10(Risk.mat[, 1]), xlim = c(-7, -4),
    ecdf.col = "green", xlab = "Risk", xaxt = "n", main = "")

> axis(1, at = -7:-4, labels = paste("1e", -7:-4, sep = ""))

> ecdfPlot(log10(Risk.mat[, 2]), ecdf.col = "blue",
    add = TRUE)

> ecdfPlot(log10(Risk.mat[, 3]), ecdf.col = "red",
    add = TRUE)

> usr <- par("usr")
```

```
> legend(x = usr[1], y = 0.9, legend = paste("Case", 1:3),
    col = c("green", "blue", "red"), lty = 1, lwd = 3,
    bty = "n")

> abline(h = 0.95, lty = 3)

> text(x = -6.5, y = 0.97, "95th %'ile")

> bounds <- apply(Risk.mat[, c(1, 3)], 2, quantile,
    probs = 0.95)

> log10.bounds <- log10(bounds)

> segments(x0 = log10.bounds, x1 = log10.bounds,
    y0 = 0, y1 = 0.95, lty = 2)

> text(x = log10.bounds, y = usr[3]/2, signif(bounds, 2))

> arrows(x0 = log10.bounds[1], x1 = log10.bounds[2],
    y0 = 0.2, y1 = 0.2, code = 3)

> text(x = mean(log10.bounds), y = 0.1,
    "Range of\nUncertainty")
```

To create Fig. 9.11, type these commands:

```
> Risk.mat.4 <- matrix(as.numeric(NA), nrow = 2000,
    ncol = 250)

> ED.max <- simulateVector(250, distribution = "unif",
    param.list = list(min = 26, max = 40),
    sample.method = "LHS", seed = 322, sort = TRUE)

> for(j in 1:250) {
    IR <- simulateVector(2000, distribution = "unif",
    param.list = list(min = 50, max = 200),
    sample.method = "LHS")

    EF <- simulateVector(n = 2000, distribution = "tri",
    param.list = list(min = 200, mode = 250, max = 350),
    sample.method = "LHS")

    ED <- simulateVector(n = 2000,
    distribution = "lnormTruncAlt",
    param.list = list(mean = 9, cv = 10/9, min = 0,
    max = ED.max[j]), sample.method = "LHS")

    Risk.mat.4[, j] <- Risk.fcn(IR = IR, EF = EF, ED = ED)
    }
```

```
> ecdfPlot(log10(Risk.mat.4[, 1]), xlim = c(-7, -4),
   ecdf.col = 1, xlab = "Risk", xaxt = "n", main = "")

> axis(1, at = -7:-4, labels = paste("1e", -7:-4, sep = ""))

> for(j in 2:250) {
   ecdfPlot(log10(Risk.mat.4[, j]), ecdf.col = j, add = TRUE)
   }
```

See USEPA (2001), Vose (2008), and Millard et al. (2014) for more examples of probabilistic risk assessment.

9.7 Summary

- Human and ecological risk assessment involves characterizing the exposure to a toxicant for one or several populations, quantifying the relationship between exposure (dose) and health or ecological effects (response), determining the risk (probability) of a health or ecological effect given the observed level(s) of exposure, and characterizing the uncertainty associated with the estimated risk.

- In the past, estimates of risk were often based solely on setting values of the input variables (e.g., body weight, dose, etc.) to particular point estimates and producing a single point estimate of risk, with little, if any, quantification of the uncertainty associated with the estimated risk. More recently, several practitioners have advocated "probabilistic" risk assessment, in which the input variables are considered random variables, so the result of the risk assessment is a probability distribution for predicted risk or exposure.

- You can use R and ENVSTATS to perform Monte Carlo simulation and probabilistic risk assessment. ENVSTATS includes functions for both simple random sampling (SRS) and Latin Hypercube sampling (LHS), as well as for generating random vectors from arbitrary distributions with a specified rank correlation matrix.

- Most risk assessment models follow the form of Eqs. 9.1 and 9.10, where the output variable (risk or exposure) is a function of several input variables, and some or all of the input variables are considered to be random variables.

- *Uncertainty analysis* involves describing the variability or distribution of values of the output variable that is due to the collective variation in the input variables. This description usually involves graphical displays such as histograms, empirical density plots, and empirical cdf plots, as well as summary statistics such as the mean, median, standard deviation, coefficient of variation, 95th percentile, etc.

- *Sensitivity analysis* involves determining how the distribution of the output variable changes with changes in the individual input variables. It is used to identify which input variables contribute the most to the variation or uncertainty in the output variable. Sensitivity analysis is also used in a broader sense to determine how changing the distributions of the input variables and/or their assumed correlations or even changing the form of the model affects the output.

- *Variability* refers to the inherent heterogeneity of a particular variable (parameter). For example, there is natural variation in body weight and height between individuals in a given population. *Uncertainty* refers to a lack of knowledge about specific parameters, models, or factors. We can usually reduce uncertainty through further measurement or study. We cannot reduce variability, since it is inherent in the variable.

- Sensitivity analysis methods include one-at-a-time deviations from a baseline case, factorial design and response surface modeling, and Monte Carlo simulation.

- Uncertainty analysis methods include describing the empirical distribution of risk, repeating the simulation using a different set of random numbers, and using mixture distributions.

- An important point to remember is that no matter how complex the mathematical model in Eqs. 9.1 or 9.10, or how extensive the uncertainty and sensitivity analyses, there is always the question of how well the mathematical model reflects reality. The only way to attempt to answer this question is with data collected directly on the input and output variables.

References

Aitchison, J. (1955). On the Distribution of a Positive Random Variable Having a Discrete Probability Mass at the Origin. *Journal of the American Statistical Association* **50**, 901–908.

Aitchison, J., and J.A.C. Brown (1957). *The Lognormal Distribution (with special references to its uses in economics)*. Cambridge University Press, London, 176 pp.

Akritas, M., T. Ruscitti, and G.P. Patil. (1994). Statistical Analysis of Censored Environmental Data. In Patil, G.P., and C.R. Rao, eds., *Handbook of Statistics, Vol. 12: Environmental Statistics*. North-Holland, Amsterdam, Chapter 7, 221–242.

Allerhand, M. (2011). *A Tiny Handbook of R*. Springer-Verlag, New York.

Amemiya, T. (1985). *Advanced Econometrics*. Harvard University Press, Cambridge, MA, 521 pp.

Anderson, H.L. (1986*). Metropolis, Monte Carlo, and the MANIAC*. Los Alamos Science, Fall 1986.

Anderson, C.W., and W.D. Ray. (1975). Improved Maximum Likelihood Estimators for the Gamma Distribution. *Communications in Statistics* **4**, 437–448.

Arlinghaus, S.L., ed. (1996). *Practical Handbook of Spatial Statistics*. CRC Press, Boca Raton, FL, 307 pp.

ASTM. (1996). *PS 64-96: Provisional Standard Guide for Developing Appropriate Statistical Approaches for Ground-Water Detection Monitoring Programs*. American Society for Testing and Materials, West Conshohocken, PA.

Bain, L.J., and M. Engelhardt. (1991). *Statistical Analysis of Reliability and Life-Testing Models*. Marcel Dekker, New York, 496 pp.

Balakrishnan, N., and A.C. Cohen. (1991). *Order Statistics and Inference: Estimation Methods*. Academic Press, San Diego, CA, 375 pp.

Barclay's California Code of Regulations. (1991). Title 22, §66264.97 [concerning hazardous waste facilities] and Title 23, §2550.7(e)(8) [concerning solid waste facilities]. Barclay's Law Publishers, San Francisco, CA.

Barnett, V., and T. Lewis. (1995). *Outliers in Statistical Data*. Third Edition. John Wiley & Sons, New York, 584 pp.

Barnett, V., and A. O'Hagan. (1997). *Setting Environmental Standards: The Statistical Approach to Handling Uncertainty and Variation*. Chapman & Hall, London, 111 pp.

Barnett, V., and K.F. Turkman, eds. (1997). *Statistics for the Environment 3: Pollution Assessment and Control*. John Wiley & Sons, New York, 345 pp.

Barr, D.R., and T. Davidson. (1973). A Kolmogorov-Smirnov Test for Censored Samples. *Technometrics* **15**(4), 739–757.

Barry, T.M. (1996). Recommendations on the Testing and Use of Pseudo-Random Number Generators Used in Monte Carlo Analysis for Risk Assessment. *Risk Analysis* **16**, 93–105.

Bartels, R. (1982). The Rank Version of von Neumann's Ratio Test for Randomness. *Journal of the American Statistical Association* **77**(377), 40–46.

Berthouex, P.M., and I. Hau. (1991a). A Simple Rule for Judging Compliance Using Highly Censored Samples. *Research Journal of the Water Pollution Control Federation* **63**(6), 880–886.

Berthouex, P.M., and I. Hau. (1991b). Difficulties Related to Using Extreme Percentiles for Water Quality Regulations. *Research Journal of the Water Pollution Control Federation* **63**(6), 873–879.

Berthouex, P.M., and L.C. Brown. (2002). *Statistics for Environmental Engineers*. Second Edition. Lewis Publishers, Boca Raton, FL.

Bacchetti, P. (2010). Current sample size conventions: Flaws, Harms, and Alternatives. *BMC Medicine* **8**, 17–23.

Birnbaum, Z. W., and F.H. Tingey. (1951). One-Sided Confidence Contours for Probability Distribution Functions. *Annals of Mathematical Statistics* **22**, 592–596.

Blom, G. (1958). *Statistical Estimates and Transformed Beta Variables*. John Wiley & Sons, New York.

Bogen, K.T. (1995). Methods to Approximate Joint Uncertainty and Variability in Risk. *Risk Analysis* **15**(3), 411–419.

Bose, A., and N. K. Neerchal. (1996). *Unbiased Estimation of Variance Using Ranked Set Sampling*. Technical Report # 96-11, Department of Statistics, Purdue University, West Lafayette, IN.

Boswell, M. T., and G. P. Patil. (1987). A Perspective of Composite Sampling. *Communications in Statistics—Theory and Methods* **16**, 3069–3093.

Box, G.E.P., and D.R. Cox. (1964). An Analysis of Transformations (with Discussion). *Journal of the Royal Statistical Society, Series B* **26**(2), 211–252.

Box, G.E.P., W.G. Hunter, and J.S. Hunter. (1978). *Statistics for Experimenters: An Introduction to Design, Data Analysis, and Model Building.* John Wiley & Sons, New York, 653 pp.

Box, G.E.P., and G.M. Jenkins. (1976). *Time Series Analysis: Forecasting and Control.* Prentice-Hall, Englewood Cliffs, NJ, 575 pp.

Bradu, D., and Y. Mundlak. (1970). Estimation in Lognormal Linear Models. *Journal of the American Statistical Association* **65**, 198–211.

Brainard, J., and D.E. Burmaster. (1992). Bivariate Distributions for Height and Weight of Men and Women in the United States. *Risk Analysis* **12**(2), 267–275.

Breslow, N.E. (1970). A Generalized Kruskal-Wallis Test for Comparing K Samples Subject to Unequal Patterns of Censorship. *Biometrika* **57**, 579–594.

Bukowski, J., L. Korn, and D. Wartenberg. (1995). Correlated Inputs in Quantitative Risk Assessment: The Effects of Distributional Shape. *Risk Analysis* **15**(2), 215–219.

Burmaster, D.E., and P.D. Anderson. (1994). Principles of Good Practice for the Use of Monte Carlo Techniques in Human Health and Ecological Risk Assessments. *Risk Analysis* **14**(4), 477–481.

Burmaster, D.E., and R.H. Harris. (1993). The Magnitude of Compounding Conservatisms in Superfund Risk Assessments. *Risk Analysis* **13**(2), 131–134.

Burmaster, D.E., and K.M. Thompson. (1997). Estimating Exposure Point Concentrations for Surface Soils for Use in Deterministic and Probabilistic Risk Assessments. *Human and Ecological Risk Assessment* **3**(3), 363–384.

Burmaster, D.E., and A.M. Wilson. (1996). An Introduction to Second-Order Random Variables in Human Health Risk Assessments. *Human and Ecological Risk Assessment* **2**(4), 892–919.

Byrd, D.M., and C.R. Cothern. (2000). *Introduction to Risk Analysis.* Government Institutes, Lanham, MD, 433 pp.

Byrnes, M.E. (1994). *Field Sampling Methods for Remedial Investigations.* Lewis Publishers, Boca Raton, FL, 254 pp.

Calitz, F. (1973). Maximum Likelihood Estimation of the Parameters of the Three-Parameter Lognormal Distribution—a Reconsideration. *Australian Journal of Statistics* **15**(3), 185–190.

Callahan, B.G., ed. (1998). Special Issue: Probabilistic Risk Assessment. *Human and Ecological Risk Assessment* **4**(2).

Callahan, B.G., D.E. Burmaster, R.L. Smith, D.D. Krewski, and B.A.F. DelloRusso, eds. (1996). Commemoration of the 50th Anniversary of Monte Carlo, *Human and Ecological Risk Assessment* **2**(4).

Castillo, E. (1988). *Extreme Value Theory in Engineering.* Academic Press, New York, 389 pp.

Castillo, E., and A. Hadi. (1994). Parameter and Quantile Estimation for the Generalized Extreme-Value Distribution. *Environmetrics* **5**, 417–432.

Caulcutt, R., and R. Boddy. (1983). *Statistics for Analytical Chemists.* Chapman & Hall, London, 253 pp.

Chambers, J.M., W.S. Cleveland, B. Kleiner, and P.A. Tukey. (1983). *Graphical Methods for Data Analysis.* Duxbury Press, Boston, MA, 395 pp.

Chandler, R.E, and E.M. Scott. (2011). *Statistical Methods for Trend Detection and Analyses in the Environmental Sciences.* John Wiley & Sons, New York.

Chen, G., and N. Balakrishnan. (1995). A General Purpose Approximate Goodness-of-Fit Test. *Journal of Quality Technology* **27**(2), 154–161.

Chen, J.J., D.W. Gaylor, and R.L. Kodell. (1990). Estimation of the Joint Risk from Multiple-Compound Exposure Based on Single-Compound Experiments. *Risk Analysis* **10**(2), 285–290.

Chen, L. (1995a). A Minimum Cost Estimator for the Mean of Positively Skewed Distributions with Applications to Estimation of Exposure to Contaminated Soils. *Environmetrics* **6**, 181–193.

Chen, L. (1995b). Testing the Mean of Skewed Distributions. *Journal of the American Statistical Association* **90**(430), 767–772.

Cheng, R.C.H., and N.A.K. Amin. (1982). Maximum Product-of-Spacings Estimation with Application to the Lognormal Distribution. *Journal of the Royal Statistical Society, Series B* **44**, 394–403.

Chew, V. (1968). Simultaneous Prediction Intervals. *Technometrics* **10**(2), 323–331.

Chou, Y.M., and D.B. Owen. (1986). One-sided Distribution-Free Simultaneous Prediction Limits for p Future Samples. *Journal of Quality Technology* **18**, 96–98.

Chowdhury, J.U., J.R. Stedinger, and L. H. Lu. (1991). Goodness-of-Fit Tests for Regional Generalized Extreme Value Flood Distributions. *Water Resources Research* **27**(7), 1765–1776.

Clark, M.J.R., and P.H. Whitfield. (1994). Conflicting Perspectives about Detection Limits and about the Censoring of Environmental Data. *Water Resources Bulletin* **30**(6), 1063–1079.

Clayton, C.A., J.W. Hines, and P.D. Elkins. (1987). Detection Limits with Specified Assurance Probabilities. *Analytical Chemistry* **59**, 2506–2514.

Cleveland, W.S. (1993). *Visualizing Data.* Hobart Press, Summit, NJ, 360 pp.

Cleveland, W.S. (1994). *The Elements of Graphing Data*. Revised Edition. Hobart Press, Summit, NJ, 297 pp.

Cochran, W.G. (1977). *Sampling Techniques*. Third Edition. John Wiley & Sons, New York, 428 pp.

Code of Federal Regulations. (1996). Definition and Procedure for the Determination of the Method Detection Limit—Revision 1.11. Title 40, Part 136, Appendix B, 7-1-96 Edition, pp. 265–267.

Cogliano, V.J. (1997). Plausible Upper Bounds: Are Their Sums Plausible? *Risk Analysis* **17**(1), 77–84.

Cohen, A.C. (1951). Estimating Parameters of Logarithmic-Normal Distributions by Maximum Likelihood. *Journal of the American Statistical Association* **46**, 206–212.

Cohen, A.C. (1959). Simplified Estimators for the Normal Distribution When Samples are Singly Censored or Truncated. *Technometrics* **1**(3), 217–237.

Cohen, A.C. (1963). Progressively Censored Samples in Life Testing. *Technometrics* **5**, 327–339.

Cohen, A.C. (1988). Three-Parameter Estimation. In Crow, E.L., and K. Shimizu, eds. *Lognormal Distributions: Theory and Applications*. Marcel Dekker, New York, Chapter 4.

Cohen, A.C. (1991). *Truncated and Censored Samples*. Marcel Dekker, New York, 312 pp.

Cohen, A.C., and B.J. Whitten. (1980). Estimation in the Three-Parameter Lognormal Distribution. *Journal of the American Statistical Association* **75**, 399–404.

Cohen, A.C., B.J. Whitten, and Y. Ding. (1985). Modified Moment Estimation for the Three-Parameter Lognormal Distribution. *Journal of Quality Technology* **17**, 92–99.

Cohen, M.A., and P.B. Ryan. (1989). Observations Less Than the Analytical Limit of Detection: A New Approach. *JAPCA* **39**(3), 328–329.

Cohn, T.A. (1988). *Adjusted Maximum Likelihood Estimation of the Moments of Lognormal Populations form Type I Censored Samples*. U.S. Geological Survey Open-File Report 88-350, 34 pp.

Cohn, T.A., L.L. DeLong, E.J. Gilroy, R.M. Hirsch, and D.K. Wells. (1989). Estimating Constituent Loads. *Water Resources Research* **25**(5), 937–942.

Cohn, T.A., E.J. Gilroy, and W.G. Baier. (1992). Estimating Fluvial Transport of Trace Constituents Using a Regression Model with Data Subject to Censoring. *Proceedings of the American Statistical Association, Section on Statistics and the Environment*.

Conover, W.J. (1980). *Practical Nonparametric Statistics*. Second Edition. John Wiley & Sons, New York, 493 pp.

Conover, W.J., M.E. Johnson, and M.M. Johnson. (1981). A Comparative Study of Tests for Homogeneity of Variances, with Applications to the Outer Continental Shelf Bidding Data. *Technometrics* **23**(4), 351–361.

Cothern, C.R. (1994). How Does Scientific Information in General and Statistical Information in Particular Input to the Environmental Regulatory Process? In Patil, G.P., and C.R. Rao, eds., *Handbook of Statistics, Vol. 12: Environmental Statistics*. North-Holland, Amsterdam, Chapter 25, 791–816.

Cothern, C.R. (1996). *Handbook for Environmental Risk Decision Making: Values, Perceptions, & Ethics*. Lewis Publishers, Boca Raton, FL, 408 pp.

Cothern, C.R., and N.P. Ross. (1994). *Environmental Statistics, Assessment, and Forecasting*. Lewis Publishers, Boca Raton, FL, 418 pp.

Cox, D.R. (1972). Regression Models and Life Tables (with Discussion). *Journal of the Royal Statistical Society of London, Series B* **34**, 187–220.

Cox, D.D., L.H. Cox, and K.B. Ensor. (1997). Spatial Sampling and the Environment: Some Issues and Directions. *Environmental and Ecological Statistics* **4**, 219–233.

Cox, D.R., and D.V. Hinkley. (1974). *Theoretical Statistics*. Chapman & Hall, New York, 511 pp.

Cox, L.H., and W.W. Piegorsch. (1996). Combining Environmental Information, I: Environmental Monitoring, Measurement and Assessment. *Environmetrics* **7**, 299–308.

Cressie, N.A.C. (1993). *Statistics for Spatial Data*. Revised Edition. John Wiley & Sons, New York, 900 pp.

Crick, M.J., M.D. Hill, and D. Charles. (1987). The role of sensitivity analysis in assessing uncertainty. In: *Proceedings of an NEA workshop on uncertainty analysis for performance assessments of radioactive waste disposal systems*, Paris, OECD, 1–258.

Crow, E.L., and K. Shimizu, eds. (1988). *Lognormal Distributions: Theory and Applications*. Marcel Dekker, New York, 387 pp.

Csuros, M. (1997). *Environmental Sampling and Analysis, Lab Manual*. Lewis Publishers, Boca Raton, FL, 373 pp.

Cullen, A.C. (1994). Measures of Compounding Conservatism in Probabilistic Risk Assessment. *Risk Analysis* **14**(4), 389–393.

Cunnane, C. (1978). Unbiased Plotting Positions—A Review. *Journal of Hydrology* **37**(3/4), 205–222.

Currie, L.A. (1968). Limits for Qualitative Detection and Quantitative Determination: Application to Radiochemistry. *Annals of Chemistry* **40**, 586–593.

Currie, L.A. (1988). *Detection in Analytical Chemistry: Importance, Theory, and Practice.* American Chemical Society, Washington, D.C.

Currie, L.A. (1995). Nomenclature in Evaluation of Analytical Methods Including Detection and Quantification Capabilities. *Pure & Applied Chemistry* **67**(10), 1699–1723.

Currie, L.A. (1996). Foundations and Future of Detection and Quantification Limits. *Proceedings of the Section on Statistics and the Environment,* American Statistical Association, Alexandria, VA.

Currie, L.A. (1997). Detection: International Update, and Some Emerging Di-Lemmas Involving Calibration, the Blank, and Multiple Detection Decisions. *Chemometrics and Intelligent Laboratory Systems* **37**, 151–181.

D'Agostino, R.B. (1970). Transformation to Normality of the Null Distribution of g1. *Biometrika* **57**, 679–681.

D'Agostino, R.B. (1971). An Omnibus Test of Normality for Moderate and Large Size Samples. *Biometrika* **58**, 341–348.

D'Agostino, R.B. (1986a). Graphical Analysis. In D'Agostino, R.B., and M.A. Stephens, eds. *Goodness-of-Fit Techniques.* Marcel Dekker, New York, Chapter 2, pp. 7–62.

D'Agostino, R.B. (1986b). Tests for the Normal Distribution. In D'Agostino, R.B., and M.A. Stephens, eds. *Goodness-of-Fit Techniques.* Marcel Dekker, New York, Chapter 9, pp. 367–419.

D'Agostino, R.B., and E.S. Pearson (1973). Tests for Departures from Normality. Empirical Results for the Distributions of $\beta2$ and $\beta1$. *Biometrika* **60**(3), 613–622.

D'Agostino, R.B., and M.A. Stephens, eds. (1986). *Goodness-of-Fit Techniques.* Marcel Dekker, New York, 560 pp.

D'Agostino, R.B., and G.L. Tietjen (1973). Approaches to the Null Distribution of Öb1. *Biometrika* **60**(1), 169–173.

Dallal, G. E., and L. Wilkinson. (1986). An Analytic Approximation to the Distribution of Lilliefor's Test for Normality. *The American Statistician* **40**, 294–296.

Danziger, L., and S. Davis. (1964). Tables of Distribution-Free Tolerance Limits. *Annals of Mathematical Statistics* **35**(5), 1361–1365.

David, M. (1977). *Geostatistical Ore Reserve Estimation.* Elsevier, New York, 364 pp.

Davidson, J.R. (1994). *Elipgrid-PC: A PC Program for Calculating Hot Spot Probabilities.* ORNL/TM-12774. Oak Ridge National Laboratory, Oak Ridge, TN.

Davis, C.B. (1994). Environmental Regulatory Statistics. In Patil, G.P., and C.R. Rao, eds., *Handbook of Statistics, Vol. 12: Environmental Statistics.* North-Holland, Amsterdam, Chapter 26, 817–865.

Davis, C.B. (1997). Challenges in Regulatory Environmetrics. *Chemometrics and Intelligent Laboratory Systems* **37**, 43–53.

Davis, C.B. (1998a). *Ground-Water Statistics & Regulations: Principles, Progress and Problems.* Second Edition. Environmetrics & Statistics Limited, Henderson, NV.

Davis, C.B. (1998b). Personal Communication, September 3, 1998.

Davis, C.B. (1998c). *Power Comparisons for Control Chart and Prediction Limit Procedures.* EnviroStat Technical Report 98-1, Environmetrics & Statistics Limited, Henderson, NV.

Davis, C.B., and R.J. McNichols. (1987). One-sided Intervals for at Least p of m Observations from a Normal Population on Each of r Future Occasions. *Technometrics* **29**, 359–370.

Davis, C.B., and R.J. McNichols. (1994a). Ground Water Monitoring Statistics Update: Part I: Progress Since 1988. *Ground Water Monitoring and Remediation* **14**(4), 148–158.

Davis, C.B., and R.J. McNichols. (1994b). Ground Water Monitoring Statistics Update: Part II: Nonparametric Prediction Limits. *Ground Water Monitoring and Remediation* **14**(4), 159–175.

Davis, C.B., and R.J. McNichols. (1999). Simultaneous Nonparametric Prediction Limits (with Discusson). *Technometrics* **41**(2), 89–112.

Davis, C.S., and M.A. Stephens. (1978). Algorithm AS 128. Approximating the Covariance Matrix of Normal Order Statistics. *Applied Statistics* **27**, 206–212.

Diggle, P.J. (1983). *Statistical Analysis of Spatial Point Patterns.* Academic Press, London.

Downing, D.J., R.H. Gardner, and F.O. Hoffman. (1985). An Examination of Response-Surface Methodologies for Uncertainty Analysis in Assessment Models. *Technometrics* **27**(2), 151–163. See also Easterling, R.G. (1986). Discussion of Downing, Gardner, and Hoffman (1985). *Technometrics* **28**(1), 91–93.

Downton, F. (1966). Linear Estimates of Parameters in the Extreme Value Distribution. *Technometrics* **8**(1), 3–17.

Draper, N., and H. Smith. (1998). *Applied Regression Analysis*. Third Edition. John Wiley & Sons, New York.

Dunnett, C.W. (1955). A Multiple Comparisons Procedure for Comparing Several Treatments with a Control. *Journal of the American Statistical Association* **50**, 1096–1121.

Dunnett, C.W. (1964). New Tables for Multiple Comparisons with a Control. *Biometrics* **20**, 482–491.

Dunnett, C.W., and M. Sobel. (1955). Approximations to the Probability Integral and Certain Percentage Points of a Multivariate Analogue of Student's t-Distribution. *Biometrika* **42**, 258–260.

Easterling, R.G. (1986). Letters to the Editor: Discussion of Downing, Gardener and Hoffman (1985). *Technometrics* **28**(1), 91–92.

Edland, S.D., and G. van Belle. (1994). Decreased Sampling Costs and Improved Accuracy with Composite Sampling. In Cothern, C.R., and N.P. Ross, eds., *Environmental Statistics, Assessment, and Forecasting*, Lewis Publishers, Boca Raton, FL, Chapter 2, pp. 29–55.

Efron, B. (1979a). Bootstrap Methods: Another Look at the Jackknife. *Annals of Statistics* **7**, 1–26.

Efron, B. (1979b). Computers and the Theory of Statistics: Thinking the Unthinkable. *Society for Industrial and Applied Mathematics* **21**, 460–480.

Efron, B. (1981). Nonparametric Standard Errors and Confidence Intervals. *Canadian Journal of Statistics* **9**, 139–172.

Efron, B. (1982). *The Jackknife, the Bootstrap and Other Resampling Plans*. Society for Industrial and Applied Mathematics, Philadelphia, PA, 92 pp.

Efron, B., and R.J. Tibshirani. (1993). *An Introduction to the Bootstrap*. Chapman & Hall, New York, 436 pp.

Einax, J.W., H.W. Zwanziger, and S. Geib. (1997). *Chemometrics in Environmental Analysis*. VCH Publishers, John Wiley & Sons, New York, 384 pp.

Ekstrøm, C.T. (2012). *The R Primer*. CRC Press, Taylor & Francis Group, Boca Raton, FL, 287 pp.

Ellison, B.E. (1964). On Two-Sided Tolerance Intervals for a Normal Distribution. *Annals of Mathematical Statistics* **35**, 762–772.

El-Shaarawi, A.H. (1989). Inferences about the Mean from Censored Water Quality Data. *Water Resources Research* **25**(4) 685–690.

El-Shaarawi, A.H. (1993). Environmental Monitoring, Assessment, and Prediction of Change. *Environmetrics* **4**(4), 381–398.

El-Shaarawi, A.H., and D.M. Dolan. (1989). Maximum Likelihood Estimation of Water Quality Concentrations from Censored Data. *Canadian Journal of Fisheries and Aquatic Sciences* **46**, 1033–1039.

El-Shaarawi, A.H., and S.R. Esterby. (1992). Replacement of Censored Observations by a Constant: An Evaluation. *Water Research* **26**(6), 835–844.

El-Shaarawi, A.H., and A. Naderi. (1991). Statistical Inference from Multiply Censored Environmental Data. *Environmental Monitoring and Assessment* **17**, 339–347.

Eschenroeder, A.Q., and E.J. Faeder. (1988). A Monte Carlo Analysis of Health Risks from PCB-Contaminated Mineral Oil Transformer Fires. *Risk Analysis* **8**(2), 291–297.

Esterby, S.R. (1993). Trend Analysis Methods for Environmental Data. *Environmetrics* **4**(4), 459–481.

Evans, M., N. Hastings, and B. Peacock. (1993). *Statistical Distributions*. Second Edition. John Wiley & Sons, New York, 170 pp.

Everitt, B.S. (2008). *Chance Rules: An Informal Guide to Probability, Risk, and Statistics*. Second Edition. Springer Science+Business Media, New York, 148 pp.

Everitt, B.S. and A. Skrondal (2006). *The Cambridge Dictionary of Statistics*. Third Edition. Cambridge University Press, New York, 478 pp.

Fertig, K.W., and N.R. Mann. (1977). One-Sided Prediction Intervals for at Least p out of m Future Observations from a Normal Population. *Technometrics* **19**, 167–177.

Fill, H.D., and J.R. Stedinger. (1995). *L* Moment and Probability Plot Correlation Coefficient Goodness-of-Fit Tests for the Gumbel Distribution and Impact of Autocorrelation. *Water Resources Research* **31**(1), 225–229.

Filliben, J.J. (1975). The Probability Plot Correlation Coefficient Test for Normality. *Technometrics* **17**(1), 111–117.

Finley, B., and D. Paustenbach. (1994). The Benefits of Probabilistic Exposure Assessment: Three Case Studies Involving Contaminated Air, Water, and Soil. *Risk Analysis* **14**(1), 53–73.

Finley, B., D. Proctor, P. Scott, N. Harrington, D. Paustenback, and P. Price. (1994). Recommended Distributions for Exposure Factors Frequently Used in Health Risk Assessment. *Risk Analysis* **14**(4), 533–553.

Finney, D.J. (1941). On the Distribution of a Variate Whose Logarithm is Normally Distributed. *Supplement to the Journal of the Royal Statistical Society* **7**, 155–161.

Fisher, L.D., and G. van Belle. (1993). *Biostatistics: A Methodology for the Health Sciences*. John Wiley & Sons, New York, 991 pp.

Fleiss, J. L. (1981). *Statistical Methods for Rates and Proportions*. Second Edition. John Wiley & Sons, New York, 321 pp.

Fleming, T.R., and D.P. Harrington. (1981). A Class of Hypothesis Tests for One and Two Sample Censored Survival Data. *Communications in Statistics— Theory and Methods* **A10**(8), 763–794.

Fleming, T.R., and D.P. Harrington. (1991). *Counting Processes and Survival Analysis*. John Wiley & Sons, New York, 429 pp.

Fletcher, D.J., L. Kavalieris, and B.F.J. Manly, eds. (1998). *Statistics in Ecology and Environmental Monitoring 2: Decision Making and Risk Assessment in Biology*. Otago Conference Series No. 6, University of Otago Press, Dunedin, New Zealand, 220 pp.

Fletcher, D.J., and B.F.J. Manly, eds. (1994). *Statistics in Ecology and Environmental Monitoring*. Otago Conference Series No. 2. University of Otago Press, Dunedin, New Zealand, 269 pp.

Flora, J.D. (1991). Statistics and Environmental Regulations. *Environmetrics* **2**(2), 129–137.

Freudenburg, W.R., and J.A. Rursch. (1994). The Risks of "Putting the Numbers in Context": A Cautionary Tale. *Risk Analysis* **14**(6), 949–958.

Gaylor, D.W., and J.J. Chen. (1996). A Simple Upper Limit for the Sum of the Risks of the Components in a Mixture. *Risk Analysis* **16**(3), 395–398.

Gehan, E.A. (1965). A Generalized Wilcoxon Test for Comparing Arbitrarily Singly-Censored Samples. *Biometrika* **52**, 203–223.

Gentle, J.E. (1985). Monte Carlo Methods. In Kotz, S., and N.L. Johnson, eds. *Encyclopedia of Statistics, Volume 5*. John Wiley & Sons, New York, 612–617.

Gibbons, R.D. (1987a). Statistical Prediction Intervals for the Evaluation of Ground-Water Quality. *Ground Water* **25**, 455–465.

Gibbons, R.D. (1987b). Statistical Models for the Analysis of Volatile Organic Compounds in Waste Disposal Sites. *Ground Water* **25**, 572–580.

Gibbons, R.D. (1990). A General Statistical Procedure for Ground-Water Detection Monitoring at Waste Disposal Facilities. *Ground Water* **28**, 235–243.

Gibbons, R.D. (1991a). Some Additional Nonparametric Prediction Limits for Ground-Water Detection Monitoring at Waste Disposal Facilities. *Ground Water* **29**, 729–736.

Gibbons, R.D. (1991b). Statistical Tolerance Limits for Ground-Water Monitoring. *Ground Water* **29**, 563–570.

Gibbons, R.D. (1994). *Statistical Methods for Groundwater Monitoring*. John Wiley & Sons, New York, 286 pp.

Gibbons, R.D. (1995). Some Statistical and Conceptual Issues in the Detection of Low-Level Environmental Pollutants (with Discussion). *Environmetrics* **2**, 125–167.

Gibbons, R.D. (1996). Some Conceptual and Statistical Issues in Analysis of Groundwater Monitoring Data. *Environmetrics* **7**, 185–199.

Gibbons, R.D., and J. Baker. (1991). The Properties of Various Statistical Prediction Intervals for Ground-Water Detection Monitoring. *Journal of Environmental Science and Health* **A26**(4), 535–553.

Gibbons, R.D., D.K. Bhaumik, and S. Aryal. (2009). *Statistical Methods for Groundwater Monitoring.* Second Edition. John Wiley & Sons, Hoboken.

Gibbons, R.D., D.E. Coleman, and R.F. Maddalone. (1997a). An Alternative Minimum Level Definition for Analytical Quantification. *Environmental Science and Technology* **31**(7), 2071–2077. Comments and Discussion in Volume 31(12), 3727–3731, and Volume 32(15), 2346–2353.

Gibbons, R.D., D.E. Coleman, and R.F. Maddalone. (1997b). Response to Comment on "An Alternative Minimum Level Definition for Analytical Quantification." *Environmental Science and Technology* **31**(12), 3729–3731.

Gibbons, R.D., D.E. Coleman, and R.F. Maddalone. (1998). Response to Comment on "An Alternative Minimum Level Definition for Analytical Quantification." *Environmental Science and Technology* **32**(15), 2349–2353.

Gibbons, R.D., N.E. Grams, F.H. Jarke, and K.P. Stoub. (1992). Practical Quantitation Limits. *Chemometrics Intelligent Laboratory Systems* **12**, 225–235.

Gibbons, R.D., F.H. Jarke, and K.P. Stoub. (1991). Detection Limits: For Linear Calibration Curves with Increasing Variance and Multiple Future Detection Decisions. In Tatsch, D.E., ed. *Waste Testing and Quality Assurance: Volume 3.* American Society for Testing and Materials, Philadelphia, PA.

Gibrat, R. (1930). Une loi des repartition economiques: l'effect proportionnel. *Bull. Statist. Gen. Fr.* **19**, 469ff.

Gilbert, R.O. (1987). *Statistical Methods for Environmental Pollution Monitoring.* Van Nostrand Reinhold, New York, 320 pp.

Gilbert, R.O. (1994). An Overview of Statistical Issues Related to Environmental Cleanup. In Patil, G.P., and C.R. Rao, eds., *Handbook of Statistics, Vol. 12: Environmental Statistics.* North-Holland, Amsterdam, Chapter 27, 867–880.

Gilliom, R.J., and D.R. Helsel. (1986). Estimation of Distributional Parameters for Censored Trace Level Water Quality Data: 1. Estimation Techniques. *Water Resources Research* **22**, 135–146.

Gilliom, R.J., R.M. Hirsch, and E.J. Gilroy. (1984). Effect of Censoring Trace-Level Water-Quality Data on Trend-Detection Capability. *Environmental Science and Technology* **18**, 530–535.

Gilroy, E.J., R.M. Hirsch, and T.A. Cohn. (1990). Mean Square Error of Regression-Based Constituent Transport Estimates. *Water Resources Research* **26**(9), 2069–2077.

Ginevan, M.E., and D.E. Splitstone. (1997). Improving Remediation Decisions at Hazardous Waste Sites with Risk-Based Geostatistical Analysis. *Environmental Science and Technology* **31**(2), 92A–96A.

Glasser, J.A., D.L. Foerst, G.D. McKee, S.A. Quave, and W.L. Budde. (1981). Trace Analyses for Wastewaters. *Environmental Science and Technology* **15**, 1426–1435.

Gleit, A. (1985). Estimation for Small Normal Data Sets with Detection Limits. *Environmental Science and Technology* **19**, 1201–1206.

Graham, R.C. (1993). *Data Analysis for the Chemical Sciences: A Guide to Statistical Techniques*. VCH Publishers, New York, 536 pp. Distributed by John Wiley & Sons, New York.

Graham, S.L., K.J. Davis, W.H. Hansen, and C.H. Graham. (1975). Effects of Prolonged Ethylene Thiourea Ingestion on the Thyroid of the Rat. *Food and Cosmetics Toxicology* **13**(5), 493–499.

Greenwood, J.A., J.M. Landwehr, N.C. Matalas, and J.R. Wallis. (1979). Probability Weighted Moments: Definition and Relation to Parameters of Several Distributions Expressible in Inverse Form. *Water Resources Research* **15**(5), 1049–1054.

Grice, J.V., and L.J. Bain. (1980). Inferences Concerning the Mean of the Gamma Distribution. *Journal of the American Statistician* **75**, 929–933.

Griffiths, D.A. (1980). Interval Estimation for the Three-Parameter Lognormal Distribution via the Likelihood Function. *Applied Statistics* **29**, 58–68.

Gringorten, I.I. (1963). A Plotting Rule for Extreme Probability Paper. *Journal of Geophysical Research* **68**(3), 813–814.

Gupta, A.K. (1952). Estimation of the Mean and Standard Deviation of a Normal Population from a Censored Sample. *Biometrika* **39**, 260–273.

Guttman, I. (1970). *Statistical Tolerance Regions: Classical and Bayesian*. Hafner Publishing Co., Darien, CT, 150 pp.

Haas, C.N. (1997). Importance of Distributional Form in Characterizing Inputs to Monte Carlo Risk Assessments. *Risk Analysis* **17**(1), 107–113.

Haas, C.N., and P.A. Scheff. (1990). Estimation of Averages in Truncated Samples. *Environmental Science and Technology* **24**(6), 912–919.

Hahn, G., and W. Nelson. (1973). A Survey of Prediction Intervals and Their Applications. *Journal of Quality Technology* **5**, 178–188.

Hahn, G.J. (1969). Factors for Calculating Two-Sided Prediction Intervals for Samples from a Normal Distribution. *Journal of the American Statistical Association* **64**(327), 878–898.

Hahn, G.J. (1970a). Additional Factors for Calculating Prediction Intervals for Samples from a Normal Distribution. *Journal of the American Statistical Association* **65**(332), 1668–1676.

Hahn, G.J. (1970b). Statistical Intervals for a Normal Population, Part I: Tables, Examples and Applications. *Journal of Quality Technology* **2**(3), 115–125.

Hahn, G.J. (1970c). Statistical Intervals for a Normal Population, Part II: Formulas, Assumptions, Some Derivations. *Journal of Quality Technology* **2**(4), 195–206.

Hahn, G.J., and W.Q. Meeker. (1991). *Statistical Intervals: A Guide for Practitioners*. John Wiley & Sons, New York, 392 pp.

Haimes, Y.Y., T. Barry, and J.H. Lambert, eds. (1994). When and How Can You Specify a Probability Distribution When You Don't Know Much? *Risk Analysis* **14**(5), 661–706.

Hall, I.J., and R.R. Prairie. (1973). One-Sided Prediction Intervals to Contain at Least m out of k Future Observations. *Technometrics* **15**, 897–914.

Hall, I.J., R.R. Prairie, and C.K. Motlagh. (1975). Non-Parametric Prediction Intervals. *Journal of Quality Technology* **7**(3), 109–114.

Hallenbeck, W.H. (1993). *Quantitative Risk Assessment for Environmental and Occupational Health*. Second Edition. Lewis Publishers, Boca Raton, FL, 224 pp.

Hamby, D.M. (1994). A Review of Techniques for Parameter Sensitivity Analysis of Environmental Models. *Environmental Monitoring and Assessment* **32**, 135–154.

Hamed, M., and P.B. Bedient. (1997). On the Effect of Probability Distributions of Input Variables in Public Health Risk Assessment. *Risk Analysis* **17**(1), 97–105.

Harrington, D.P., and T.R. Fleming. (1982). A Class of Rank Test Procedures for Censored Survival Data. *Biometrika* **69**(3), 553–566.

Harter, H.L., and A.H. Moore. (1966). Local-Maximum-Likelihood Estimation of the Parameters of Three-Parameter Lognormal Populations from Complete and Censored Samples. *Journal of the American Statistical Association* **61**, 842–851.

Hashimoto, L.K., and R.R. Trussell. (1983). Evaluating Water Quality Data Near the Detection Limit. Paper presented at the Advanced Technology Conference, American Water Works Association, Las Vegas, Nevada, June 5–9, 1983.

Hastie, T.J., and R.J. Tibshirani. (1990). *Generalized Additive Models*. Chapman & Hall, New York.

Hattis, D. (1990). Three Candidate "Laws" of Uncertainty Analysis. *Risk Analysis* **10**(1), 11.

Hattis, D., and D.E. Burmaster. (1994). Assessment of Variability and Uncertainty Distributions for Practical Risk Analyses. *Risk Analysis* **14**(5), 713–730.

Hawkins, D. M., and R.A.J. Wixley. (1986). A Note on the Transformation of Chi-Squared Variables to Normality. *The American Statistician* **40**, 296–298.

Hayes, B. (1993). The Wheel of Fortune. *American Scientist* **81**, 114–118.

Hazardous Materials Training and Research Institute (HMTRI). (1997). *Site Characterization: Sampling and Analysis*. Van Nostrand Reinhold, 316 pp. Distributed by John Wiley & Sons, New York.

Helsel, D.R. (1987). Advantages of Nonparametric Procedures for Analysis of Water Quality Data. *Hydrological Sciences Journal* **32**, 179–190.

Helsel, D.R. (1990). Less than Obvious: Statistical Treatment of Data Below the Detection Limit. *Environmental Science and Technology* **24**(12), 1766–1774.

Helsel, D.R. (2012). *Statistics for Censored Environmental Data Using Minitab and R, 2nd Edition*. John Wiley & Sons, Hoboken, New Jersey, 344 pp.

Helsel, D.R., and T.A. Cohn. (1988). Estimation of Descriptive Statistics for Multiply Censored Water Quality Data. *Water Resources Research* **24**(12), 1997–2004.

Helsel, D.R., and R.J. Gilliom. (1986). Estimation of Distributional Parameters for Censored Trace Level Water Quality Data: 2. Verification and Applications. *Water Resources Research* **22**, 147–155.

Helsel, D.R., and R.M. Hirsch. (1988). Discussion of Applicability of the t-test for Detecting Trends in Water Quality Variables. *Water Resources Bulletin* **24**(1), 201–204.

Helsel, D.R., and R.M. Hirsch. (1992). *Statistical Methods in Water Resources Research*. Elsevier, New York, 522 pp.

Helsel, D.R., and R.M. Hirsch. (2002). *Statistical Methods in Water Resources Research*. Techniques of Water Resources Investigations, Book 4, Chapter A3. U.S. Geological Survey. Available on-line at http://www.practicalstats.com/aes/aes/AESbook.html.

Hettmansperger, T.P. (1984). *Statistical Inference Based on Ranks*. John Wiley & Sons, New York, 323 pp.

Heyde, C.C. (1963). On a Property of the Lognormal Distribution. *Journal of the Royal Statistical Society, Series B* **25**, 392–393.

Hill, B.M. (1963). The Three-Parameter Lognormal Distribution and Bayesian Analysis of a Point-Source Epidemic. *Journal of the American Statistical Association* **58**, 72–84.

Hinkley, D.V., and G. Runger. (1984). The Analysis of Transformed Data (with Discussion). *Journal of the American Statistical Association* **79**, 302–320.

Hirsch, R.M. (1988). Statistical Methods and Sampling Design for Estimating Step Trends in Surface-Water Quality. *Water Resources Bulletin* **24**(3), 493–503.

Hirsch, R.M., R.B. Alexander, and R.A. Smith. (1991). Selection of Methods for the Detection and Estimation of Trends in Water Quality. *Water Resources Research* **27**(5), 803–813.

Hirsch, R.M., and E.J. Gilroy. (1984). Methods of Fitting a Straight Line to Data: Examples in Water Resources. *Water Resources Bulletin* **20**(5), 705–711.

Hirsch, R.M., D.R. Helsel, T.A. Cohn, and E.J. Gilroy. (1993). Statistical Analysis of Hydrologic Data. In Maidment, D.R., ed. *Handbook of Hydrology*. McGraw-Hill, New York, Chapter 17.

Hirsch, R.M., and J.R. Slack. (1984). A Nonparametric Trend Test for Seasonal Data with Serial Dependence. *Water Resources Research* **20**(6), 727–732.

Hirsch, R.M., J.R. Slack, and R.A. Smith. (1982). Techniques of Trend Analysis for Monthly Water Quality Data. *Water Resources Research* **18**(1), 107–121.

Hirsch, R.M., and J.R. Stedinger. (1987). Plotting Positions for Historical Floods and Their Precision. *Water Resources Research* **23**(4), 715–727.

Hoaglin, D.C. (1988). Transformations in Everyday Experience. *Chance* **1**, 40–45.

Hoaglin, D.C., and D.F. Andrews. (1975). The Reporting of Computation-Based Results in Statistics. *The American Statistician* **29**(3), 122–126.

Hoaglin, D.C., F. Mosteller, and J.W. Tukey, eds. (1983). *Understanding Robust and Exploratory Data Analysis*. John Wiley & Sons, New York, 447 pp.

Hoeffding, W. (1948). A Class of Statistics with Asymptotically Normal Distribution. *Annals of Mathematical Statistics* **19**, 293–325.

Hoffman, F.O., and J.S. Hammonds. (1994). Propagation of Uncertainty in Risk Assessments: The Need to Distinguish between Uncertainty Due to Lack of Knowledge and Uncertainty Due to Variability. *Risk Analysis* **14**(5), 707–712.

Hollander, M., and D.A. Wolfe. (1999). *Nonparametric Statistical Methods*. Second Edition. John Wiley & Sons, New York, 787 pp.

Hoshi, K., J.R. Stedinger, and J. Burges. (1984). Estimation of Log-Normal Quantiles: Monte Carlo Results and First-Order Approximations. *Journal of Hydrology* **71**, 1–30.

Hosking, J.R.M. (1984). Testing Whether the Shape Parameter is Zero in the Generalized Extreme-Value Distribution. *Biometrika* **71**(2), 367–374.

Hosking, J.R.M. (1985). Algorithm AS 215: Maximum-Likelihood Estimation of the Parameters of the Generalized Extreme-Value Distribution. *Applied Statistics* **34**(3), 301–310.

Hosking, J.R.M. (1990). L-Moments: Analysis and Estimation of Distributions Using Linear Combinations of Order Statistics. *Journal of the Royal Statistical Society, Series B* **52**(1), 105–124.

Hosking, J.R.M., and J.R. Wallis (1995). A Comparison of Unbiased and Plotting-Position Estimators of L Moments. *Water Resources Research* **31**(8), 2019–2025.

Hosking, J.R.M., J.R. Wallis, and E.F. Wood. (1985). Estimation of the Generalized Extreme-Value Distribution by the Method of Probability-Weighted Moments. *Technometrics* **27**(3), 251–261.

Hosmer, D.W, S. Lemeshow, and S. May. (2008). *Applied Survival Analysis: Regression Modeling of Time to Event Data, 2nd Edition.* John Wiley & Sons, Hoboken, New Jersey, 416 pp.

Hubaux, A., and G. Vos. (1970). Decision and Detection Limits for Linear Calibration Curves. *Annals of Chemistry* **42**, 849–855.

Hughes, J.P., and S.P. Millard. (1988). A Tau-Like Test for Trend in the Presence of Multiple Censoring Points. *Water Resources Bulletin* **24**(3), 521–531.

Ibrekk, H., and M.G. Morgan. (1987). Graphical Communication of Uncertain Quantities to Nontechnical People. *Risk Analysis* **7**(4), 519–529.

Iman, R.L., and W.J. Conover. (1980). Small Sample Sensitivity Analysis Techniques for Computer Models, With an Application to Risk Assessment (with Comments). *Communications in Statistics—Volume A, Theory and Methods* **9**(17), 1749–1874.

Iman, R.L., and W.J. Conover. (1982). A Distribution-Free Approach to Inducing Rank Correlation Among Input Variables. *Communications in Statistics—Volume B, Simulation and Computation* **11**(3), 311–334.

Iman, R.L., and J.M. Davenport. (1982). Rank Correlation Plots For Use with Correlated Input Variables. *Communications in Statistics—Volume B, Simulation and Computation* **11**(3), 335–360.

Iman, R.L., and J.C. Helton. (1988). An Investigation of Uncertainty and Sensitivity Analysis Techniques for Computer Models. *Risk Analysis* **8**(1), 71–90.

Iman, R.L., and J.C. Helton. (1991). The Repeatability of Uncertainty and Sensitivity Analyses for Complex Probabilistic Risk Assessments. *Risk Analysis* **11**(4), 591–606.

Isaaks, E.H., and R.M. Srivastava. (1989). *An Introduction to Applied Geostatistics*. Oxford University Press, New York, 561 pp.

Israeli, M., and C.B. Nelson. (1992). Distribution and Expected Time of Residence for U.S. Households. *Risk Analysis* **12**(1), 65–72.

Jenkinson, A.F. (1955). The Frequency Distribution of the Annual Maximum (or Minimum) of Meteorological Events. *Quarterly Journal of the Royal Meteorological Society* **81**, 158–171.

Jenkinson, A.F. (1969). *Statistics of Extremes*. Technical Note 98, World Meteorological Office, Geneva.

Johnson, G.D., B.D. Nussbaum, G.P. Patil, and N.P. Ross. (1996). Designing Cost-Effective Environmental Sampling Using Concomitant Information. *Chance* **9**(1), 4–11.

Johnson, N.L., S. Kotz, and N. Balakrishnan. (1994). *Continuous Univariate Distributions, Volume 1*. Second Edition. John Wiley & Sons, New York, 756 pp.

Johnson, N.L., S. Kotz, and N. Balakrishnan. (1995). *Continuous Univariate Distributions, Volume 2*. Second Edition. John Wiley & Sons, New York, 719 pp.

Johnson, N.L., S. Kotz, and N. Balakrishnan. (1997). *Discrete Multivariate Distributions*. John Wiley & Sons, New York, 299 pp.

Johnson, N.L., S. Kotz, and A.W. Kemp. (1992). *Univariate Discrete Distributions*. Second Edition. John Wiley & Sons, New York, 565 pp.

Johnson, N.L., and B.L. Welch. (1940). Applications of the Non-Central t-Distribution. *Biometrika* **31**, 362–389.

Johnson, R.A., D.R. Gan, and P.M. Berthouex. (1995). Goodness-of-Fit Using Very Small but Related Samples with Application to Censored Data Estimation of PCB Contamination. *Environmetrics* **6**, 341–348.

Johnson, R.A., S. Verrill, and D.H. Moore. (1987). Two-Sample Rank Tests for Detecting Changes That Occur in a Small Proportion of the Treated Population. *Biometrics* **43**, 641–655.

Johnson, R.A., and D.W. Wichern. (2007). *Applied Multivariate Statistical Analysis*. Sixth Edition. Prentice-Hall, Englewood Cliffs, NJ.

Journel, A.G., and Huijbregts, C.J. (1978). *Mining Geostatistics*. Academic Press, London.

Judge, G.G., W.E. Griffiths, R.C. Hill, H. Lutkepohl, and T.C. Lee. (1985). Qualitative and Limited Dependent Variable Models: Chapter 18. In *The Theory and Practice of Econometrics*. John Wiley & Sons, New York, 1019 pp.

Kahn, H.D., W.A. Telliard, and C.E. White. (1998). Comment on "An Alternative Minimum Level Definition for Analytical Quantification" (with Response). *Environmental Science and Technology* **32**(5), 2346–2353.

Kahn, H.D., C.E. White, K. Stralka, and R. Kuznetsovski. (1997). Alternative Estimates of Detection. *Proceedings of the Twentieth Annual EPA Conference on Analysis of Pollutants in the Environment*, May 7–8, Norfolk, VA. U.S. Environmental Protection Agency, Washington, D.C.

Kaiser, H. (1965). Zum Problem der Nachweisgrenze. *Fresenius' Z. Anal. Chem.* **209**, 1.

Kalbfleisch, J.D., and R.L. Prentice. (1980). *The Statistical Analysis of Failure Time Data*. John Wiley & Sons, New York, 321 pp.

Kallioras, A.G., I.A. Koutrouvelis, and G.C. Canavos. (2006). Testing the Fit of a Gamma Distribution Using the Empirical Moment Generating Function. *Communications in Statistics – Theory and Methods* **35**, 527–540.

Kaluzny, S.P., S.C. Vega, T.P. Cardosa, and A.A. Shelly. (1998). *S+SPATIALSTATS User's Manual for Windows and UNIX*. Springer-Verlag, New York, 327 pp.

Kaplan, E.L., and P. Meier. (1958). Nonparametric Estimation from Incomplete Observations. *Journal of the American Statistical Association* **53**, 457–481.

Kapteyn, J.C. (1903). *Skew Frequency Curves in Biology and Statistics*. Astronomical Laboratory, Noordhoff, Groningen.

Keith, L.H. (1991). *Environmental Sampling and Analysis: A Practical Guide*. Lewis Publishers, Boca Raton, FL, 143 pp.

Keith, L.H., ed. (1996). *Principles of Environmental Sampling*. Second Edition. American Chemical Society, Washington, D.C., 848 pp. Distributed by Oxford University Press, New York.

Kempthorne, O., and T.E. Doerfler. (1969). The Behavior of Some Significance Tests under Experimental Randomization. *Biometrika* **56**(2), 231–248.

Kendall, M.G., and J.D. Gibbons. (1990). *Rank Correlation Methods*. Fifth Edition. Charles Griffin, London, 260 pp.

Kendall, M.G., and A. Stuart. (1991). *The Advanced Theory of Statistics, Volume 2: Inference and Relationship*. Fifth Edition. Oxford University Press, New York, 1323 pp.

Kennedy, W.J., and J.E. Gentle. (1980). *Statistical Computing*. Marcel Dekker, New York, 591 pp.

Kim, P.J., and R.I. Jennrich. (1973). Tables of the Exact Sampling Distribution of the Two Sample Kolmogorov-Smirnov Criterion. In Harter, H.L., and D.B. Owen, eds. *Selected Tables in Mathematical Statistics, Vol. 1*. American Mathematical Society, Providence, RI, pp. 79–170.

Kimbrough, D.E. (1997). Comment on "An Alternative Minimum Level Definition for Analytical Quantification" (with Response). *Environmental Science and Technology* **31**(12), 3727–3731.

Kitanidis, P.K. (1997). *Introduction to Geostatistics*. Cambridge University Press, London, 249 pp.

Kleinbaum, D.G., and M. Klein (2011). *Survival Analysis: A Self-Learning Text, Third Edition*. Springer, New York, 700 pp.

Kodell, R.L., H. Ahn, J.J. Chen, J.A. Springer, C.N. Barton, and R.C. Hertzberg. (1995). Upper Bound Risk Estimates for Mixtures of Carcinogens. *Toxicology* **105**, 199–208.

Kodell, R.L., and J.J. Chen. (1994). Reducing Conservatism in Risk Estimation for Mixtures of Carcinogens. *Risk Analysis* **14**(3), 327–332.

Kolmogorov, A.N. (1933). Sulla determinazione empirica di una legge di distribuzione. *Giornale dell' Istituto Italiano degle Attuari* **4**, 83–91.

Korn, L.R., and D.E. Tyler. (2001). Robust Estimation for Chemical Concentration Data Subject to Detection Limits. In Fernholz, L., S. Morgenthaler, and W. Stahel, eds. *Statistics in Genetics and in the Environmental Sciences*. Birkhauser Verlag, Basel, pp.41–63.

Kotz, S., and N.L. Johnson, eds. (1985). *Encyclopedia of Statistics*. John Wiley & Sons, New York.

Krishnamoorthy K., and T. Mathew. (2009). *Statistical Tolerance Regions: Theory, Applications, and Computation*. John Wiley and Sons, Hoboken.

Krishnamoorthy, K., T. Matthew, and G. Ramachandaran. (2006). Generalized P-values and Confidence Intervals: a Novel Approach for Analyzing Lognormally Distributed Exposure Data. *Journal of Occupational and Environmental Hygene* **3**, 642–650.

Krishnamoorthy K., T. Mathew, and S. Mukherjee. (2008). Normal-Based Methods for a Gamma Distribution: Prediction and Tolerance Intervals and Stress-Strength Reliability. *Technometrics* **50**(1), 69–78.

Kuchenhoff, H., and M. Thamerus. (1996). Extreme Value Analysis of Munich Air Pollution Data. *Environmental and Ecological Statistics* **3**, 127–141.

Kulkarni, H.V., and S.K. Powar. (2010). A New Method for Interval Estimation of the Mean of the Gamma Distribution. *Lifetime Data Analysis* **16**, 431–447.

Kushner, E.J. (1976). On Determining the Statistical Parameters for Pollution Concentrations from a Truncated Data Set. *Atmospheric Environment* **10**, 975–979.

Lambert, D., B. Peterson, and I. Terpenning. (1991). Nondetects, Detection Limits, and the Probability of Detection. *Journal of the American Statistical Association* **86**(414), 266–277.

Lambert, J.H., N.C. Matalas, C.W. Ling, Y.Y. Haimes, and D. Li. (1994). Selection of Probability Distributions in Characterizing Risk of Extreme Events. *Risk Analysis* **14**(5), 731–742.

Land, C.E. (1971). Confidence Intervals for Linear Functions of the Normal Mean and Variance. *The Annals of Mathematical Statistics* **42**(4), 1187–1205.

Land, C.E. (1972). An Evaluation of Approximate Confidence Interval Estimation Methods for Lognormal Means. *Technometrics* **14**(1), 145–158.

Land, C.E. (1973). Standard Confidence Limits for Linear Functions of the Normal Mean and Variance. *Journal of the American Statistical Association* **68**(344), 960–963.

Land, C.E. (1975). Tables of confidence limits for linear functions of the normal mean and variance, in *Selected Tables in Mathematical Statistics, Vol. III.* American Mathematical Society, Providence, RI, pp. 385–419.

Landwehr, J.M., N.C. Matalas, and J.R. Wallis. (1979). Probability Weighted Moments Compared with Some Traditional Techniques in Estimating Gumbel Parameters and Quantiles. *Water Resources Research* **15**(5), 1055–1064.

Laudan, L. (1997). *Danger Ahead: The Risks You Really Face on Life's Highway.* John Wiley & Sons, New York, 203 pp.

Law, A.M. (2006). *Simulation Modeling & Analysis.* Fourth Edition. McGraw-Hill, Inc., New York, 792 pp.

Latta, R.B. (1981). A Monte Carlo Study of Some Two-Sample Rank Tests with Censored Data. *Journal of the American Statistical Association* **76**(375), 713–719.

Lee, E.T., and J.W. Wang (2003). *Statistical Methods for Survival Data Analysis.* Third Edition. John Wiley & Sons, Hoboken, New Jersey, 513 pp.

Lehmann, E.L. (1975). *Nonparametrics: Statistical Methods Based on Ranks.* Holden-Day, Oakland, CA, 457 pp.

Leidel, N.A., K.A. Busch, and J.R. Lynch. (1977). *Occupational Exposure Sampling Strategy Manual.* U.S. Department of Health, Education, and Welfare, Public Health Service, Center for Disease Control, National Institute for Occupational Safety and Health, Cincinnati, OH, January, 1977.

Lewis, H.W. (1990). *Technological Risk.* W.W. Norton & Company, New York, 353 pp.

Lettenmaier, D.P. (1976). Detection of Trends in Water Quality Data from Data with Dependent Observations. *Water Resources Research* **12**, 1037–1046.

Lettenmaier, D.P. (1988). Multivariate Non-Parametric Tests for Trend in Water Quality. *Water Resources Bulletin* **24**(3), 505–512.

Likes, J. (1980). Variance of the MVUE for Lognormal Variance. *Technometrics* **22**(2), 253–258.

Looney, S.W., and T.R. Gulledge. (1985). Use of the Correlation Coefficient with Normal Probability Plots. *The American Statistician* **39**(1), 75–79.

Lovison, G., S.D. Gore, and G.P. Patil. (1994). Design and Analysis of Composite Sampling Procedures: A Review. In Patil, G.P., and C.R. Rao, eds., *Handbook of Statistics, Vol. 12: Environmental Statistics*. North-Holland, Amsterdam, Chapter 4, 103–166.

Lu, L., and J.R. Stedinger. (1992). Variance of Two- and Three-Parameter GEV/PWM Quantile Estimators: Formulae, Confidence Intervals, and a Comparison. *Journal of Hydrology* **138**, 247–267.

Lucas, J.M. (1982). Combined Shewart-CUMSUM Quality Control Schemes. *Journal of Quality Technology* **14**, 51–59.

Lucas, J.M. (1985). Cumulative Sum (CUMSUM) Control Schemes. *Communications in Statistics A—Theory and Methods* **14**(11), 2689–2704.

Lundgren, R., and A. McMakin. (2009). *Risk Communication: A Handbook for Communicating Environmental, Safety, and Health Risks*. Fourth Edition. IEEE Press, John Wiley & Sons, Hoboken, NJ, 380 pp.

Macleod, A.J. (1989). Remark AS R76: A Remark on Algorithm AS 215: Maximum Likelihood Estimation of the Parameters of the Generalized Extreme-Value Distribution. *Applied Statistics* **38**(1), 198–199.

Manly, B.F.J. (1997). Randomization, Bootstrap and Monte Carlo Methods in Biology. Second Edition. Chapman & Hall, New York, 399 pp.

Mann, H.B. (1945). Nonparametric Tests Against Trend. *Econometrica* **13**, 245–259.

Mantel, N. (1966). Evaluation of Survival Data and Two New Rank Order Statistics Arising in its Consideration. *Cancer Chemotherapy Reports* **50**, 163–170.

Maritz, J.S., and A.H. Munro. (1967). On the Use of the Generalized Extreme-Value Distribution in Estimating Extreme Percentiles. *Biometrics* **23**(1), 79–103.

Marsaglia, G. et al. (1973). *Random Number Package: "Super-Duper."* School of Computer Science, McGill University, Montreal.

Massart, D.L., B.G.M. Vandeginste, S.N. Deming, Y. Michotte, and L. Kaufman. (1988). *Chemometrics: A Textbook*. Elsevier, New York, Chapter 7.

Matsumoto, M. and Nishimura, T. (1998). Mersenne Twister: A 623-Dimensionally Equidistributed Uniform Pseudo-Random Number Generator. *ACM Transactions on Modeling and Computer Simulation*, **8**, 3–30.

McBean, E.A., and F.A. Rovers. (1992). Estimation of the Probability of Exceedance of Contaminant Concentrations. *Ground Water Monitoring Review* **Winter**, 115–119.

McBean, E.A., and R.A. Rovers. (1998). *Statistical Procedures for Analysis of Environmental Monitoring Data & Risk Assessment.* Prentice-Hall PTR, Upper Saddle River, NJ, 313 pp.

McCullagh, P., and J.A. Nelder. (1989). *Generalized Linear Models.* Second Edition. Chapman & Hall, New York, 511 pp.

McIntyre, G.A. (1952). A Method of Unbiased Selective Sampling, Using Ranked Sets. *Australian Journal of Agricultural Resources* **3**, 385–390.

McKay, M.D., R.J. Beckman., and W.J. Conover. (1979). A Comparison of Three Methods for Selecting Values of Input Variables in the Analysis of Output from a Computer Code. *Technometrics* **21**(2), 239–245.

McKean, J.W., and G.L. Sievers. (1989). Rank Scores Suitable for Analysis of Linear Models under Asymmetric Error Distributions. *Technometrics* **31**, 207–218.

McKone, T.E. (1994). Uncertainty and Variability in Human Exposures to Soil and Contaminants through Home-Grown Food: A Monte Carlo Assessment. *Risk Analysis* **14**(4), 449–463.

McKone, T.E., and K.T. Bogen. (1991). Predicting the Uncertainties in Risk Assessment. *Environmental Science and Technology* **25**(10), 1674–1681.

McLeod, A.I., K.W. Hipel, and B.A. Bodo. (1991). Trend Analysis Methodology for Water Quality Time Series. *Environmetrics* **2**(2), 169–200.

McNichols, R.J., and C.B. Davis. (1988). Statistical Issues and Problems in Ground Water Detection Monitoring at Hazardous Waste Facilities. *Ground Water Monitoring Review* **8**(4), 135–150.

Meeker, W.Q., and L.A. Escobar. (1998). *Statistical Methods for Reliability Data.* John Wiley & Sons, New York.

Meier, P.C., and R.E. Zund. (1993). *Statistical Methods in Analytical Chemistry.* John Wiley & Sons, New York, 321 pp.

Michael, J.R., and W.R. Schucany. (1986). Analysis of Data from Censored Samples. In D'Agostino, R.B., and M.A. Stephens, eds. *Goodness-of-Fit Techniques.* Marcel Dekker, New York, 560 pp., Chapter 11, 461–496.

Millard, S.P. (1987a). Environmental Monitoring, Statistics, and the Law: Room for Improvement (with Comment). *The American Statistician* **41**(4), 249–259.

Millard, S.P. (1987b). Proof of Safety vs. Proof of Hazard. *Biometrics* **43**, 719–725.

Millard, S.P. (1996). Estimating a Percent Reduction in Load. *Water Resources Research* **32**(6), 1761–1766.

Millard, S.P. (1998). *ENVIRONMENTALSTATS for S-PLUS User's Manual for Windows and UNIX*. Springer-Verlag, New York, 381 pp.

Millard, S.P., P. Dixon, and N.K. Neerchal. (2014). *Environmental Statistics with R*. CRC Press, Boca Raton, Florida.

Millard, S.P., and S.J. Deverel. (1988). Nonparametric Statistical Methods for Comparing Two Sites Based on Data with Multiple Nondetect Limits. *Water Resources Research* **24**(12), 2087–2098.

Millard, S.P., and D.P. Lettenmaier. (1986). Optimal Design of Biological Sampling Programs Using the Analysis of Variance. *Estuarine, Coastal and Shelf Science* **22**, 637–656.

Millard, S.P., and N.K. Neerchal. (2001). *Environmental Statistics with S-PLUS*. CRC Press, Boca Raton, Florida, 830 pp.

Millard, S.P., J.R. Yearsley, and D.P. Lettenmaier. (1985). Space-Time Correlation and Its Effects on Methods for Detecting Aquatic Ecological Change. *Canadian Journal of Fisheries and Aquatic Science*, **42**(8), 1391–1400. Correction: (1986), 43, 1680.

Miller, R.G. (1981a). *Simultaneous Statistical Inference*. Springer-Verlag, New York, 299 pp.

Miller, R.G. (1981b). *Survival Analysis*. John Wiley & Sons, New York, 238 pp.

Milliken, G.A., and D.E. Johnson. (1992). *Analysis of Messy Data, Volume I: Designed Experiments*. Chapman & Hall, New York, 473 pp.

Mode, N.A., L.L. Conquest, and D.A. Marker. (1999). Ranked Set Sampling for Ecological Research: Accounting for the Total Costs of Sampling. *Environmetrics* **10**, 179–194.

Montgomery, D.C. (1997). *Introduction to Statistical Quality Control*. Third Edition. John Wiley & Sons, New York.

Moore, D.S. (1986). Tests of Chi-Squared Type. In D'Agostino, R.B., and M.A. Stephens, eds. *Goodness-of-Fit Techniques*. Marcel Dekker, New York, pp. 63–95.

Morgan, M.G., and M. Henrion. (1990). *Uncertainty: A Guide to Dealing with Uncertainty in Quantitative Risk and Policy Analysis*. Cambridge University Press, New York, 332 pp.

Neely, W.B. (1994). *Introduction to Chemical Exposure and Risk Assessment*. Lewis Publishers, Boca Raton, FL, 190 pp.

Neerchal, N. K., and S. L. Brunenmeister. (1993). Estimation of Trend in Chesapeake Bay Water Quality Data. In Patil, G.P., and C.R. Rao, eds., *Handbook of Statistics, Vol. 6: Multivariate Environmental Statistics*. North-Holland, Amsterdam, Chapter 19, 407–422.

Nelson, W. (1969). Hazard Plotting for Incomplete Failure Data. *Journal of Quality Technology* **1**(1), 27–52.

Nelson, W. (1972). Theory and Applications of Hazard Plotting for Censored Failure Data. *Technometrics* **14**, 945–966.

Nelson, W. (1982). *Applied Life Data Analysis*. John Wiley & Sons, New York, 634 pp.

Nelson, W. (2004). *Accelerated Testing: Statistical Models, Test Plans, and Data Analysis*. John Wiley & Sons, Hoboken, New Jersey,

Nelson, W., and J. Schmee. (1979). Inference for (Log) Normal Life Distributions from Small Singly Censored Samples and BLUEs. *Technometrics* **21**, 43–54.

Nelson, W.R. (1970). Confidence Intervals for the Ratio of Two Poisson Means and Poisson Predictor Intervals. *IEEE Transactions on Reliability* **R-19**(2), 42–49.

Neptune, D., E.P. Brantly, M.J. Messner, and D.I. Michael. (1990). Quantitative Decision Making in Superfund: A Data Quality Objectives Case Study. *Hazardous Material Control* **May/June**, pp. 18–27.

Newman, M.C. (1995). *Quantitative Methods in Aquatic Ecotoxicology*. Lewis Publishers, Boca Raton, FL, 426 pp.

Newman, M.C., and P.M. Dixon. (1990). UNCENSOR: A Program to Estimate Means and Standard Deviations for Data Sets with Below Detection Limit Observations. *American Environmental Laboratory* **4(90)**, 26–30.

Newman, M.C., P.M. Dixon, B.B. Looney, and J.E. Pinder. (1989). Estimating Mean and Variance for Environmental Samples with Below Detection Limit Observations. *Water Resources Bulletin* **25**(4), 905–916.

Odeh, R.E., and D.B. Owen. (1980). *Tables for Normal Tolerance Limits, Sampling Plans, and Screening*. Marcel Dekker, New York, 316 pp.

Osborne, C. (1991). Statistical Calibration: A Review. *International Statistical Review* **59**(3), 309–336.

Ostrom, L.T., and C. Wilhelmsen. (2012). *Risk Assessment: Tools, Techniques, and Their Applications*. John Wiley & Sons, Hoboken, 416 pp.

Ott, W.R. (1990). A Physical Explanation of the Lognormality of Pollutant Concentrations. *Journal of the Air and Waste Management Association* **40**, 1378–1383.

Ott, W.R. (1995). *Environmental Statistics and Data Analysis*. Lewis Publishers, Boca Raton, FL, 313 pp.

Owen, D.B. (1962). *Handbook of Statistical Tables*. Addison-Wesley, Reading, MA, 580 pp.

Owen, W., and T. DeRouen. (1980). Estimation of the Mean for Lognormal Data Containing Zeros and Left-Censored Values, with Applications to the Measurement of Worker Exposure to Air Contaminants. *Biometrics* **36**, 707–719.

Parkin, T.B., S.T. Chester, and J.A. Robinson. (1990). Calculating Confidence Intervals for the Mean of a Lognormally Distributed Variable. *Journal of the Soil Science Society of America* **54**, 321–326.

Parkin, T.B., J.J. Meisinger, S.T. Chester, J.L. Starr, and J.A. Robinson. (1988). Evaluation of Statistical Estimation Methods for Lognormally Distributed Variables. *Journal of the Soil Science Society of America* **52**, 323–329.

Parmar, M.K.B., and D. Machin. (1995). *Survival Analysis: A Practical Approach*. John Wiley & Sons, New York, 255 pp.

Patil, G.P., S.D. Gore, and A.K. Sinha. (1994). Environmental Chemistry, Statistical Modeling, and Observational Economy. In Cothern, C.R., and N.P. Ross, eds., *Environmental Statistics, Assessment, and Forecasting*, Lewis Publishers, Boca Raton, FL, Chapter 3, pp. 57–97.

Patil, G.P., and C.R. Rao, eds. (1994). *Handbook of Statistics, Vol. 12: Environmental Statistics*. North-Holland, Amsterdam, 927 pp.

Patil, G.P., A.K. Sinha, and C. Taillie. (1994). Ranked Set Sampling. In Patil, G.P., and C.R. Rao, eds., *Handbook of Statistics, Vol. 12: Environmental Statistics*. North-Holland, Amsterdam, Chapter 5, 167–200.

Pearson, E.S., and H.O. Hartley, eds. (1970). *Biometrika Tables for Statisticians, Volume 1*. Cambridge University Press, New York, 270 pp.

Persson, T. and H. Rootzen. (1977). Simple and Highly Efficient Estimators for a Type I Censored Normal Sample. *Biometrika* **64**, 123–128.

Peterson, B.P., S.P. Millard, E.F. Wood, and D.P. Lettenmaier. (1990). Design of a Soil Sampling Study to Determine the Habitability of the Emergency Declaration Area, Love Canal, New York. *Environmetrics* **1**(1), 89–119.

Peto, R., and J. Peto. (1972). Asymptotically Efficient Rank Invariant Test Procedures (with Discussion). *Journal of the Royal Statistical Society of London, Series A* **135**, 185–206.

Pettitt, A.N. (1976). Cramér-von Mises Statistics for Testing Normality with Censored Samples. *Biometrika* **63**(3), 475–481.

Pettitt, A. N. (1983). Re-Weighted Least Squares Estimation with Censored and Grouped Data: An Application of the EM Algorithm. *Journal of the Royal Statistical Society, Series B* **47**, 253–260.

Pettitt, A.N., and M.A. Stephens. (1976). Modified Cramér-von Mises Statistics for Censored Data. *Biometrika* **63**(2), 291–298.

Piegorsch, W.W., and A.J. Bailer. (1997). *Statistics for Environmental Biology and Toxicology*. Chapman & Hall, New York.

Piegorsch, W.W., and A.J. Bailer. (2005). *Analyzing Environmental Data*. John Wiley & Sons, New York.

Pomeranz, J. (1973). Exact Cumulative Distribution of the Kolmogorov-Smirnov Statistic for Small Samples (Algorithm 487). *Collected Algorithms from CACM*.

Porter, P.S., R.C. Ward, and H.F. Bell. (1988). The Detection Limit. *Environmental Science and Technology* **22**(8), 856–861.

Prentice, R.L. (1978). Linear Rank Tests with Right Censored Data. *Biometrika* **65**, 167–179.

Prentice, R.L. (1985). Linear Rank Tests. In Kotz, S., and N.L. Johnson, eds. *Encyclopedia of Statistical Science*. John Wiley & Sons, New York. Volume 5, pp. 51–58.

Prentice, R.L., and P. Marek. (1979). A Qualitative Discrepancy between Censored Data Rank Tests. *Biometrics* **35**, 861–867.

Prescott, P., and A.T. Walden. (1980). Maximum Likelihood Estimation of the Parameters of the Generalized Extreme-Value Distribution. *Biometrika* **67**(3), 723–724.

Prescott, P., and A.T. Walden. (1983). Maximum Likelihood Estimation of the Three-Parameter Generalized Extreme-Value Distribution from Censored Samples. *Journal of Statistical Computing and Simulation* **16**, 241–250.

Press, W.H., S.A. Teukolsky, W.T. Vetterling, and B.P. Flannery. (1992). *Numerical Recipes in FORTRAN: The Art of Scientific Computing*. Second Edition. Cambridge University Press, New York, 963 pp.

Qian, S.S. (1997). Estimating the Area Affected by Phosphorus Runoff in an Everglades Wetland: A Comparison of Universal Kriging and Bayesian Kriging. *Environmental and Ecological Statistics* **4**, 1–29.

Ranasinghe, J.A., L.C. Scott, and R. Newport. (1992). *Long-term Benthic Monitoring and Assessment Program for the Maryland Portion of the Bay, Jul 1984-Dec 1991*. Report prepared for the Maryland Department of the Environment and the Maryland Department of Natural Resources by Versar, Inc., Columbia, MD.

Ripley, B.D. (1981). *Spatial Statistics*. John Wiley & Sons, New York.

Robson, M.G., and W.A. Toscano, eds. (2007). *Risk Assessment for Environmental Health*. Jossey-Bass, San Francisco, CA, 664 pp.

Rocke, D.M., and S. Lorenzato. (1995). A Two-Component Model for Measurement Error in Analytical Chemistry. *Technometrics* **37**(2), 176–184.

Rodricks, J.V. (2007). *Calculated Risks: The Toxicity and Human Health Risks of Chemicals in Our Environment*. Second Edition. Cambridge University Press, New York, 358 pp.

Rogers, J. (1992). Assessing Attainment of Ground Water Cleanup Standards Using Modified Sequential t-Tests. *Environmetrics* **3**(3), 335–359.

Rosebury, A.M., and D.E. Burmaster. (1992). Lognormal Distributions for Water Intake by Children and Adults. *Risk Analysis* **12**(1), 99–104.

Rowe, W.D. (1994). Understanding Uncertainty. *Risk Analysis* **14**(5), 743–750.

Royston, J.P. (1982). Algorithm AS 177. Expected Normal Order Statistics (Exact and Approximate). *Applied Statistics* **31**, 161–165.

Royston, J.P. (1992a). Approximating the Shapiro-Wilk W-Test for Non-Normality. *Statistics and Computing* **2**, 117–119.

Royston, J.P. (1992b). Estimation, Reference Ranges and Goodness of Fit for the Three-Parameter Log-Normal Distribution. *Statistics in Medicine* **11**, 897–912.

Royston, J.P. (1992c). A Pocket-Calculator Algorithm for the Shapiro-Francia Test of Non-Normality: An Application to Medicine. *Statistics in Medicine* **12**, 181–184.

Royston, P. (1993). A Toolkit for Testing for Non-Normality in Complete and Censored Samples. *The Statistician* **42**, 37–43.

Rubinstein, R.Y. (1981). *Simulation and the Monte Carlo Method*. John Wiley & Sons, New York, 278 pp.

Ruffle, B., D.E. Burmaster, P.D. Anderson, and H.D. Gordon. (1994). Lognormal Distributions for Fish Consumption by the General U.S. Population. *Risk Analysis* **14**(4), 395–404.

Rugen, P., and B. Callahan. (1996). An Overview of Monte Carlo, A Fifty Year Perspective. *Human and Ecological Risk Assessment* **2**(4), 671–680.

Ryan, T.P. (1989). *Statistical Methods for Quality Improvement*. John Wiley & Sons, New York, 446 pp.

Saltelli, A., and J. Marivoet. (1990). Non-Parametric Statistics in Sensitivity Analysis for Model Output: A Comparison of Selected Techniques. *Reliability Engineering and System Safety* **28**, 229–253.

Sara, Martin N. (1994). *Standard Handbook for Solid and Hazardous Waste Facility Assessment*. Lewis Publishers, Boca Raton, FL. Chapter 11.

Sarhan, A.E., and B.G. Greenberg. (1956). Estimation of Location and Scale Parameters by Order Statistics for Singly and Doubly Censored Samples, Part I, The Normal Distribution up to Samples of Size 10. *Annals of Mathematical Statistics* **27**, 427–457.

Saw, J.G. (1961a). Estimation of the Normal Population Parameters Given a Type I Censored Sample. *Biometrika* **48**, 367–377.

Saw, J.G. (1961b). The Bias of the Maximum Likelihood Estimators of Location and Scale Parameters Given a Type II Censored Normal Sample. *Biometrika* **48**, 448–451.

Schaeffer, D.J., H.W. Kerster, and K.G. Janardan. (1980). Grab Versus Composite Sampling: A Primer for Managers and Engineers. *Journal of Environmental Management* **4**, 157–163.

Scheffé, H. (1959). *The Analysis of Variance*. John Wiley & Sons, New York, 477 pp.

Scheuer, E.M., and D.S. Stoller. (1962). On the Generation of Normal Random Vectors. *Technometrics* **4**(2), 278–281.

Schmee, J., D.Gladstein, and W. Nelson. (1985). Confidence Limits for Parameters of a Normal Distribution from Singly Censored Samples, Using Maximum Likelihood. *Technometrics* **27**(2), 119–128.

Schmoyer, R.L., J.J. Beauchamp, C.C. Brandt, and F.O. Hoffman, Jr. (1996). Difficulties with the Lognormal Model in Mean Estimation and Testing. *Environmental and Ecological Statistics* **3**, 81–97.

Schneider, H. (1986). *Truncated and Censored Samples from Normal Populations*. Marcel Dekker, New York, 273 pp.

Schneider, H., and L. Weisfeld. (1986). Inference Based on Type II Censored Samples. *Biometrics* **42**, 531–536.

Seiler, F.A. (1987). Error Propagation for Large Errors. *Risk Analysis* **7**(4), 509–518.

Seiler, F.A., and J.L. Alvarez. (1996). On the Selection of Distributions for Stochastic Variables. *Risk Analysis* **16**(1), 5–18.

Sengupta, S., and N. K. Neerchal. (1994). Classification of Grab Samples Using Composite Sampling: An Improved Strategy. *Calcutta Statistical Association Bulletin* **44**, 195–208.

Serfling, R.J. (1980). *Approximation Theorems of Mathematical Statistics*. John Wiley & Sons, New York, 371 pp.

Shaffner, G. (1999). *The Arithmetic of Life*. Ballantine Books, New York, 208 pp.

Shapiro, S.S., and C.W. Brian. (1981). A Review of Distributional Testing Procedures and Development of a Censored Sample Distributional Test. In Taillie, C., G.P. Patil, and B.A. Baldessari, eds. *Statistical Distributions in Scientific Work, Volume 5—Inferential Problems and Properties*, 1–24.

Shapiro, S.S., and R.S. Francia. (1972). An Approximate Analysis of Variance Test for Normality. *Journal of the American Statistical Association* **67**(337), 215–219.

Shapiro, S.S., and M.B. Wilk. (1965). An Analysis of Variance Test for Normality (Complete Samples). *Biometrika* **52**, 591–611.

Shea, B.L., and A.J. Scallon. (1988). Remark AS R72. A Remark on Algorithm AS 128. Approximating the Covariance Matrix of Normal Order Statistics. *Applied Statistics* **37**, 151–155.

Sheskin, D.J. (1997). *Handbook of Parametric and Nonparametric Statistical Procedures*. CRC Press, Boca Raton, FL, 719 pp.

Shlyakhter, A.I. (1994). An Improved Framework for Uncertainty Analysis: Accounting for Unsuspected Errors. *Risk Analysis* **14**(4), 441–447.

Shumway, R.H., A.S. Azari, and P. Johnson. (1989). Estimating Mean Concentrations under Transformations for Environmental Data with Detection Limits. *Technometrics* **31**(3), 347–356.

Silverman, B. W. (1986). *Density Estimation for Statistics and Data Analysis*. Chapman & Hall, New York.

Singh, A. (1993). Multivariate Decision and Detection Limits. *Analytica Chimica Acta* **277**, 205–214.

Singh, A.K., A. Singh, and M. Engelhardt. (1997). *The Lognormal Distribution in Environmental Applications*. EPA/600/R-97/006. December, 1997. Technology Support Center for Monitoring and Site Characterization, Technology Innovation Office, Office of Research and Development, Office of Solid Waste and Emergency Response, U.S. Environmental Protection Agency, Washington, D.C.

Singh, A., and J. Nocerino. (2002). Robust Estimation of Mean and Variance Using Environmental Data Sets with Below Detection Limit Observations. *Chemometrics and Intelligent Laboratory Systems* **60**, 69– 86.

Singh, A., A.K. Singh, and R.J. Iaci. (2002). Estimation of the Exposure Point Concentration Term Using a Gamma Distribution. EPA/600/R-02/084. October 2002. Technology Support Center for Monitoring and Site Characterization, Office of Research and Development, Office of Solid Waste and Emergency Response, U.S. Environmental Protection Agency, Washington, D.C.

Singh, A., R. Maichle, and S. Lee. (2006). On the Computation of a 95% Upper Confidence Limit of the Unknown Population Mean Based Upon Data Sets with Below Detection Limit Observations. EPA/600/R-06/022, March 2006. Office of Research and Development, U.S. Environmental Protection Agency, Washington, D.C.

Singh, A., R. Maichle, and N. Armbya. (2010a). *ProUCL Version 4.1.00 User Guide (Draft)*. EPA/600/R-07/041, May 2010. Office of Research and Development, U.S. Environmental Protection Agency, Washington, D.C.

Singh, A., N. Armbya, and A. Singh. (2010b). *ProUCL Version 4.1.00 Technical Guide (Draft)*. EPA/600/R-07/041, May 2010. Office of Research and Development, U.S. Environmental Protection Agency, Washington, D.C.

Size, W.B., ed. (1987). *Use and Abuse of Statistical Methods in the Earth Sciences*. Oxford University Press, New York, 169 pp.

Slob, W. (1994). Uncertainty Analysis in Multiplicative Models. *Risk Analysis* **14**(4), 571–576.

Smirnov, N.V. (1939). Estimate of Deviation between Empirical Distribution Functions in Two Independent Samples. *Bulletin Moscow University* **2**(2), 3–16.

Smirnov, N.V. (1948). Table for Estimating the Goodness of Fit of Empirical Distributions. *Annals of Mathematical Statistics* **19**, 279–281.

Smith, A.E., P.B. Ryan, and J.S. Evans. (1992). The Effect of Neglecting Correlations When Propagating Uncertainty and Estimating the Population Distribution of Risk. *Risk Analysis* **12**(4), 467–474.

Smith, E.P. (1994). Biological Monitoring: Statistical Issues and Models. In Patil, G.P., and C.R. Rao, eds., *Handbook of Statistics, Vol. 12: Environmental Statistics*. North-Holland, Amsterdam, Chapter 8, 243–261.

Smith, E.P., and K. Rose. (1991). Trend Detection in the Presence of Covariates: Stagewise Versus Multiple Regression. *Environmetrics* **2**(2), 153–168.

Smith, R.L. (1985). Maximum Likelihood Estimation in a Class of Nonregular Cases. *Biometrika* **72**(1), 67–90.

Smith, R.L. (1994). Use of Monte Carlo Simulation for Human Exposure Assessment at a Superfund Site. *Risk Analysis* **14**(4), 433–439.

Smith, R.M., and L.J. Bain. (1976). Correlation Type Goodness-of-Fit Statistics with Censored Sampling. *Communications in Statistics A—Theory and Methods* **5**(2), 119–132.

Snedecor, G.W., and W.G. Cochran. (1989). *Statistical Methods*. Eighth Edition. Iowa State University Press, Ames, IA, 503 pp.

Sokal, R.F., and F.J. Rolfe. (1981). *Biometry: The Principals and Practice of Statistics in Biological Research*. Second Edition. W.H. Freeman and Company, San Francisco, 859 pp.

Spiegelman, C.H. (1997). A Discussion of Issues Raised by Lloyd Currie and a Cross Disciplinary View of Detection Limits and Estimating Parameters That Are Often at or near Zero. *Chemometrics and Intelligent Laboratory Systems* **37**, 183–188.

Springer, M.D. (1979). *The Algebra of Random Variables*. John Wiley & Sons, New York, 470 pp.

Starks, T.H. (1988). *Evaluation of Control Chart Methodologies for RCRA Waste Sites*. Draft Report by Environmental Research Center, University of Nevada, Las Vegas, for Exposure Assessment Research Division, Environmental Monitoring Systems Laboratory—Las Vegas, Nevada. EPA Technical Report CR814342-01-3.

Stedinger, J. (1983). Confidence Intervals for Design Events. *Journal of Hydraulic Engineering* **109**(1), 13–27.

Stedinger, J.R. (1980). Fitting Lognormal Distributions to Hydrologic Data. *Water Resources Research* **16**(3), 481–490.

Stedinger, J.R., R.M. Vogel, and E. Foufoula-Georgiou. (1993). Frequency Analysis of Extreme Events. In Maidment, D.R., ed. *Handbook of Hydrology*. McGraw-Hill, New York, Chapter 18.

Stein, M. (1987). Large Sample Properties of Simulations Using Latin Hypercube Sampling. *Technometrics* **29**(2), 143–151.

Stephens, M.A. (1970). Use of the Kolmogorov-Smirnov, Cramér-von Mises and Related Statistics Without Extensive Tables. *Journal of the Royal Statistical Society, Series B* **32**, 115–122.

Stephens, M.A. (1986a). Tests Based on EDF Statistics. In D'Agostino, R. B., and M.A. Stevens, eds. *Goodness-of-Fit Techniques*. Marcel Dekker, New York, Chapter 4, pp. 97–193.

Stephens, M.A. (1986b). Tests Based on Regression and Correlation. In D'Agostino, R. B., and M.A. Stevens, eds. *Goodness-of-Fit Techniques*. Marcel Dekker, New York, Chapter 5, pp. 195–233.

Stokes, S.L. (1980). Estimation of Variance Using Judgment Ordered Ranked Set Samples. *Biometrics* **36**, 35–42.

Stolarski, R.S., A.J. Krueger, M.R. Schoeberl, R.D. McPeters, P.A. Newman, and J.C. Alpert. (1986). Nimbus 7 Satellite Measurements of the Springtime Antarctic Ozone Decrease. *Nature* **322**, 808–811.

Stoline, M.R. (1991). An Examination of the Lognormal and Box and Cox Family of Transformations in Fitting Environmental Data. *Environmetrics* **2**(1), 85–106.

Stoline, M.R. (1993). Comparison of Two Medians Using a Two-Sample Lognormal Model in Environmental Contexts. *Environmetrics* **4**(3), 323–339.

Suter, G.W. (2007). *Ecological Risk Assessment*. Second Edition. CRC Press, Boca Raton, FL, 680 pp.

Suter, G.W., R.A. Efroymson, B.E. Sample, and D.S. Jones (2000). *Ecological Risk Assessment for Contaminated Sites*. CRC Press, Boca Raton, FL, 680 pp.

Taylor, J.K. (1987). *Quality Assurance of Chemical Measures*. Lewis Publishers, CRC Press, Boca Raton, FL, 328 pp.

Taylor, J.K. (1990). *Statistical Techniques for Data Analysis*. Lewis Publishers, Boca Raton, FL, 200 pp.

Thompson, K.M., and D.E. Burmaster. (1991). Parametric Distributions for Soil Ingestion by Children. *Risk Analysis* **11**(2), 339–342.

Thompson, K.M., D.E. Burmaster, and E.A.C. Crouch. (1992). Monte Carlo Techniques for Quantitative Uncertainty Analysis in Public Health Risk Assessments. *Risk Analysis* **12**(1), 53–63.

Tiago de Oliveira, J. (1963). Decision Results for the Parameters of the Extreme Value (Gumbel) Distribution Based on the Mean and Standard Deviation. *Trabajos de Estadistica* **14**, 61–81.

Tiago de Oliveira, J. (1983). Gumbel Distribution. In Kotz, S., and N. Johnson, eds. *Encyclopedia of Statistical Sciences*. John Wiley & Sons, New York. Volume 3, pp. 552–558.

Travis, C.C., and M.L. Land. (1990). Estimating the Mean of Data Sets with Nondetectable Values. *Environmental Science and Technology* **24**, 961–962.

Tsui, K.L., N.P. Jewell, and C.F.J. Wu. (1988). A Nonparametric Approach to the Truncated Regression Problem. *Journal of the American Statistical Association* **83**(403), 785–792.

USEPA. (1983). Standard for Remedial Actions at Inactive Uranium Processing Sites; Final Rule (40 CFR Part 192). *Federal Register* **48**(3), 590–604.

USEPA. (1987a). *Data Quality Objectives for Remedial Response Activities, Development Process*. EPA/540/G-87/003. U.S. Environmental Protection Agency, Washington, D.C.

USEPA. (1987b). *Data Quality Objectives for Remedial Response Activities, Example Scenario RI/FS Activities at a Site with Contaminated Soils and Ground Water.* EPA/540/G-87/004. U.S. Environmental Protection Agency, Washington, D.C.

USEPA. (1987c). List (Phase 1) of Hazardous Constituents for Ground-Water Monitoring; Final Rule. *Federal Register* **52**(131), 25942–25953 (July 9, 1987).

USEPA. (1988). Statistical Methods for Evaluating Ground-Water Monitoring from Hazardous Waste Facilities: Final Rule. *Federal Register* **53**, 39, 720–731.

USEPA. (1989a). *Methods for Evaluating the Attainment of Cleanup Standards, Volume 1: Soils and Solid Media.* EPA/230-02-89-042. Office of Policy, Planning, and Evaluation, U.S. Environmental Protection Agency, Washington, D.C.

USEPA. (1989b). *Statistical Analysis of Ground-Water Monitoring Data at RCRA Facilities, Interim Final Guidance.* EPA/530-SW-89-026. Office of Solid Waste, U.S. Environmental Protection Agency, Washington, D.C.

USEPA. (1989c). *Risk Assessment Guidance for Superfund, Volume 1: Human Health Evaluation Manual (Part A), Interim Final Guidance.* EPA/540/1-89/002. Office of Emergency and Remedial Response, U.S. Environmental Protection Agency, Washington, D.C.

USEPA. (1990). *Test Methods for Evaluating Solid Waste, 3rd Edition, Proposed Update I.* Office of Solid Waste and Emergency Response, U.S. Environmental Protection Agency, Washington, D.C.

USEPA. (1991a). *Data Quality Objectives Process for Planning Environmental Data Collection Activities, Draft.* Quality Assurance Management Staff, Office of Research and Development, U.S. Environmental Protection Agency, Washington, D.C.

USEPA. (1991b). *Risk Assessment Guidance for Superfund, Volume 1: Human Health Evaluation Manual, Supplemental Guidance/Standard Default Exposure Factors, Interim Final.* OSWER Directive 9285.6-03. Office of Emergency and Remedial Response, U.S. Environmental Protection Agency, Washington, D.C.

USEPA. (1991c). Solid Waste Disposal Facility Criteria: Final Rule. *Federal Register* **56**, 50978–51119.

USEPA. (1992a). *Characterizing Heterogeneous Wastes: Methods and Recommendations.* EPA/600/R-92/033. U.S. Environmental Protection Agency, Washington, D.C.

USEPA. (1992b). *Methods for Evaluating the Attainment of Cleanup Standards, Volume 2: Groundwater*. EPA/230-R-92-014. Office of Policy, Planning, and Evaluation, U.S. Environmental Protection Agency, Washington, D.C.

USEPA. (1992c). *Statistical Analysis of Ground-Water Monitoring Data at RCRA Facilities: Addendum to Interim Final Guidance*. Office of Solid Waste, U.S. Environmental Protection Agency, Washington, D.C. Currently available as part of: Statistical Training Course for Ground-Water Monitoring Data Analysis, EPA/530-R-93-003, which may be obtained through the RCRA Docket (202/260-9327).

USEPA. (1992d). *Supplemental Guidance to RAGS: Calculating the Concentration Term*. Publication 9285.7-081, May 1992. Intermittent Bulletin, Volume 1, Number 1. Office of Emergency and Remedial Response, Hazardous Site Evaluation Division, OS-230. Office of Solid Waste and Emergency Response, U.S. Environmental Protection Agency, Washington, D.C.

USEPA. (1992e). *Guidance for Data Useability in Risk Assessment (Part A) – Final*. Publication 9285.7-09A, PB92-963356, April 1992. Office of Emergency and Remedial Response, U.S. Environmental Protection Agency, Washington, D.C.

USEPA. (1992f). *Guidance for Data Useability in Risk Assessment (Part B) – Final*. Publication 9285.7-09B, PB92-963362, May 1992. Office of Emergency and Remedial Response, U.S. Environmental Protection Agency, Washington, D.C.

USEPA. (1992g). *Guidelines for Exposure Assessment*. EPA/600/Z-92/001. Risk Assessment Forum, U.S. Environmental Protection Agency, Washington, D.C.

USEPA. (1994a). *Guidance for the Data Quality Objectives Process, EPA QA/G-4*. EPA/600/R-96/005. Office of Research and Development, U.S. Environmental Protection Agency, Washington, D.C.

USEPA. (1994b). *Statistical Methods for Evaluating the Attainment of Cleanup Standards, Volume 3: Reference-Based Standards for Soils and Solid Media*. EPA/230-R-94-004. Office of Policy, Planning, and Evaluation, U.S. Environmental Protection Agency, Washington, D.C.

USEPA. (1994c). *Use of Monte Carlo Simulation in Risk Assessments*. EPA/903-F-94-001. Hazardous Waste Management Division, U.S. Environmental Protection Agency, Region III, Philadelphia, PA.

USEPA. (1995a). *EPA Observational Economy Series, Volume 1: Composite Sampling*. EPA/230-R-95-005. Office of Policy, Planning, and Evaluation, U.S. Environmental Protection Agency, Washington, D.C.

USEPA. (1995b). *EPA Observational Economy Series, Volume 2: Ranked Set Sampling*. EPA/230-R-95-006. Office of Policy, Planning, and Evaluation, U.S. Environmental Protection Agency, Washington, D.C.

USEPA. (1995c). *EPA Risk Characterization Policy and Guidance.* Memorandum from Carol M. Browner, Administrator, U.S. Environmental Protection Agency, March 21, 1995.

USEPA. (1996a). *Guidance for Data Quality Assessment: Practical Methods for Data Analysis, EPA QA/G-9, QA96 Version.* EPA/600/R-96/084, July 1996. Office of Research and Development, U.S. Environmental Protection Agency, Washington, D.C.

USEPA. (1996b). *Soil Screening Guidance: User's Guide.* EPA/540/R-96/018, PB96963505. Office of Emergency and Remedial Response, U.S. Environmental Protection Agency, Washington, D.C., April, 1996.

USEPA. (1996c). *Soil Screening Guidance: Technical Background Document.* EPA/540/R-95/128, PB96963502. Office of Emergency and Remedial Response, U.S. Environmental Protection Agency, Washington, D.C., May, 1996.

USEPA. (1997a). *Guiding Principles for Monte Carlo Analysis.* EPA/630/R-97/001. Risk Assessment Forum, U.S. Environmental Protection Agency, Washington, D.C.

USEPA. (1997b). *Policy for Use of Monte Carlo Analysis in Risk Assessment, With Attachment (Draft).* Memorandum from William P. Wood, January 29, 1997. U.S. Environmental Protection Agency, Washington, D.C.

USEPA. (1997c). *Ecological Risk Assessment Guidance for Superfund: Process for Designing and Conducting Ecological Risk Assessments, Interim Final.* EPA 540-R-97-006, OSWER 9285.7-25, PB97-963211, June 1997. Office of Solid Waste and Emergency Response, U.S. Environmental Protection Agency, Washington, D.C.

USEPA. (1997d). *The Lognormal Distribution in Environmental Applications.* EPA/600/S-97/006, December 1997. Office of Research and Development, and Office of Solid Waste and Emergency Response, U.S. Environmental Protection Agency, Washington, D.C.

USEPA. (1998a). *Guidance for Data Quality Assessment: Practical Methods for Data Analysis, EPA QA/G-9, QA97 Version.* EPA/600/R-96/084, January 1998. Office of Research and Development, U.S. Environmental Protection Agency, Washington, D.C.

USEPA. (1998b). *Guidance for Quality Assurance Project Plans, EPA QA/G-5.* EPA/600/R-98/018. Office of Research and Development, U.S. Environmental Protection Agency, Washington, D.C.

USEPA. (1999). *Report of the Workshop on Selecting Input Distributions For Probabilistic Assessments.* EPA/630/R-98/004. Risk Assessment Forum, U.S. Environmental Protection Agency, Washington, D.C.

USEPA. (2001). *Risk Assessment Guidance for Superfund: Volume III - Part A, Process for Conducting Probabilistic Risk Assessment*. EPA 540-R-02-002, OSWER 9285.7-45, PB2002 963302, December 2001. Office of Emergency and Remedial Response, U.S. Environmental Protection Agency, Washington, D.C.

USEPA. (2002a). *Guidance for Comparing Background and Chemical Concentrations in Soil for CERCLA Sites*. EPA 540-R-01-003, OSWER 9285.7-41, September 2002. Office of Emergency and Remedial Response, U.S. Environmental Protection Agency, Washington, D.C.

USEPA. (2002b). *Supplemental Guidance for Developing Soil Screening Levels for Superfund Sites*. OSWER 9355.4-24, December 2002. Office of Solid Waste and Emergency Response, U.S. Environmental Protection Agency, Washington, D.C.

USEPA. (2002c). *Guidance on Choosing a Sampling Design for Environmental Data Collection for Use in Developing a Quality Assurance Project Plan, EPA QA/G-5S*. EPA/240/R-02/005, December 2002. Office of Environmental Information, U.S. Environmental Protection Agency, Washington, D.C.

USEPA. (2002d). *Calculating Upper Confidence Limits for Exposure Point Concentrations at Hazardous Waste Sites*. OSWER 9285.6-10, December 2002. Office of Emergency and Remedial Response, U.S. Environmental Protection Agency, Washington, D.C.

USEPA. (2005). *Guidelines for Carcinogen Risk Assessment*. EPA/630/P-03/001F, March, 2005. Risk Assessment Forum, U.S. Environmental Protection Agency, Washington, D.C.

USEPA. (2006a). *Data Quality Assessment: Statistical Methods for Practitioners, EPA QA/G-9S*. EPA/240/B-06/003, February 2006. Office of Environmental Information, U.S. Environmental Protection Agency, Washington, D.C.

USEPA. (2006b). *Guidance on Systematic Planning Using the Data Quality Objectives Process, EPA QA/G-4*. EPA/240/B-06/001, February 2006. Office of Environmental Information, U.S. Environmental Protection Agency, Washington, D.C.

USEPA. (2006c). *Data Quality Assessment: A Reviewer's Guide, EPA QA/G-9R*. EPA/240/B-06/002, February, 2006. Office of Environmental Information, U.S. Environmental Protection Agency, Washington, D.C.

USEPA. (2007). *Systematic Planning: A Case Study of Particulate Matter Ambient Air Monitoring, EPA QA/CS-2*. EPA/240/B-07/001, March 2007. Office of Environmental Information, U.S. Environmental Protection Agency, Washington, D.C.

USEPA. (2009). *Statistical Analysis of Groundwater Monitoring Data at RCRA Facilities: Unified Guidance.* EPA 530-R-09-007, March 2009. Office of Resource Conservation and Recovery, Program Implementation and Information Division, U.S. Environmental Protection Agency, Washington, D.C.

van Belle, G. (2008). *Statistical Rules of Thumb.* Second Edition. John Wiley and Sons, New York.

van Belle, G., L.D. Fisher, P.J. Heargerty, and T. Lumley. (2004). *Biostatistics: A Methodology for the Health Sciences.* Second Edition. John Wiley & Sons, New York.

van Belle, G., and J.P. Hughes. (1984). Nonparametric Tests for Trend in Water Quality. *Water Resources Research* **20**(1), 127–136.

van Belle, G., and D.C. Martin. (1993). Sample Size as a Function of Coefficient of Variation and Ratio of Means. *The American Statistician* **47**(3), 165–167.

Venables, W.N., and B.D. Ripley. (1999). *Modern Applied Statistics with S-PLUS.* Third Edition. Springer-Verlag, New York, 501 pp.

Venzon, D.J., and S.H. Moolgavkar. (1988). A Method for Computing Profile-Likelihood-Based Confidence Intervals. *Journal of the Royal Statistical Society, Series C (Applied Statistics)* **37**(1), 87–94.

Verrill, S., and R.A. Johnson. (1987). The Asymptotic Equivalence of Some Modified Shapiro-Wilk Statistics—Complete and Censored Sample Cases. *The Annals of Statistics* **15**(1), 413–419.

Verrill, S., and R.A. Johnson. (1988). Tables and Large-Sample Distribution Theory for Censored-Data Correlation Statistics for Testing Normality. *Journal of the American Statistical Association* **83**, 1192–1197.

Vogel, R.M. (1986). The Probability Plot Correlation Coefficient Test for the Normal, Lognormal, and Gumbel Distributional Hypotheses. *Water Resources Research* **22**(4), 587–590. (Correction, *Water Resources Research* **23**(10), 2013, 1987.)

Vogel, R.M., and N.M. Fennessey. (1993). L Moment Diagrams Should Replace Product Moment Diagrams. *Water Resources Research* **29**(6), 1745–1752.

Vogel, R.M., and D.E. McMartin. (1991). Probability Plot Goodness-of-Fit and Skewness Estimation Procedures for the Pearson Type 3 Distribution. *Water Resources Research* **27**(12), 3149–3158.

Vose, D. (2008). *Risk Analysis: A Quantitative Guide.* Third Edition. John Wiley & Sons, West Sussex, UK, 752 pp.

Wald, A., and J. Wolfowitz. (1946). Tolerance Limits for a Normal Distribution. *Annals of Mathematical Statistics* **17**, 208–215.

Walsh, J. (1996). *True Odds: How Risk Affects Your Everyday Life*. Silver Lake Publishing, Aberdeen, WA, 401 pp.

Webster, R., and M.A. Oliver. (1990). *Statistical Methods in Soil and Land Resource Survey*. Oxford University Press, New York, 316 pp.

Weinstein, N.D., P.M. Sandman, and W.K. Hallman. (1994). Testing a Visual Display to Explain Small Probabilities. *Risk Analysis* **14**(6), 895–896.

Weisberg, S. (1985). *Applied Linear Regression*. Second Edition. John Wiley & Sons, New York, 324 pp.

Weisberg, S. (2005). *Applied Linear Regression*. Third Edition. John Wiley & Sons, Hoboken, 336 pp.

Weisberg, S., and C. Bingham. (1975). An Approximate Analysis of Variance Test for Non-Normality Suitable for Machine Calculation. *Technometrics* **17**(1), 133–134.

Wicksell, S.D. (1917). On Logarithmic Correlation with an Application to the Distribution of Ages at First Marriage. *Medd. Lunds. Astr. Obs.* **84**, 1–21.

Wilk, M.B., and R. Gnanadesikan. (1968). Probability Plotting Methods for the Analysis of Data. *Biometrika* **55**, 1–17.

Wilk, M.B., and S.S. Shapiro. (1968). The Joint Assessment of Normality of Several Independent Samples. *Technometrics* **10**(4), 825–839.

Wilks, S.S. (1941). Determination of Sample Sizes for Setting Tolerance Limits. *Annals of Mathematical Statistics* **12**, 91–96.

Wilson, E.B., and M.M. Hilferty. (1931). The Distribution of Chi-Squares. *Proceedings of the National Academy of Sciences* **17**, 684–688.

WSDOE. (1992). *Statistical Guidance for Ecology Site Managers*. Washington State Department of Ecology, Olympia, WA.

WSDOE. (1993). *Statistical Guidance for Ecology Site Managers, Supplement S-6: Analyzing Site or Background Data with Below-Detection Limit or Below-PQL Values (Censored Data Sets)*. Washington State Department of Ecology, Olympia, WA.

Zacks, S. (1970). Uniformly Most Accurate Upper Tolerance Limits for Monotone Likelihood Ratio Families of Discrete Distributions. *Journal of the American Statistical Association* **65**, 307–316.

Zar, J.H. (2010). *Biostatistical Analysis*. Fifth Edition. Prentice-Hall, Upper Saddle River, NJ.

Zorn, M.E., R.D. Gibbons, and W.C. Sonzogni. (1997). Weighted Least-Squares Approach to Calculating Limits of Detection and Quantification by Modeling Variability as a Function of Concentration. *Analytical Chemistry* **69**, 3069–3075.

Zou, G.Y., C.Y. Huo, and J. Taleban. (2009). Simple Confidence Intervals for Lognormal Means and their Differences with Environmental Applications. *Environmetrics* **20**, 172–180.

Zurr, A.F., E.N. Ieno, and E.H.W.G. Meesters. (2009). *A Beginner's Guide to R*. Springer, New York, 218 pp.

Index

A

Abbreviations
    probability distribution names, 79, 80, 96
Acknowledgements, vii–ix
ACL. *See* Alternate concentration limit
    (ACL)
Alternate concentration limit (ACL), 102
Analysis of variance (ANOVA), 50, 149,
    163, 165, 172
ANOVA. *See* Analysis of variance
    (ANOVA)
Assessment monitoring. *See* Compliance
    monitoring
Autocorrelation. *See* Serial correlation

B

Beta distribution, 79
Binomial distribution, 28, 55, 102
    estimating parameters for, 99
Binomial proportion
    confidence interval for, 99
        sample size for, 34
    estimating, 99
Bootstrap, 101, 187, 189, 190
Box-Cox transformations
    censored data, 183–184
    optimal, 3, 63, 76, 78, 176
    PPCC *vs.* transform power, 72
    problems with, 71
    TcCB data, 72
Boxplot, 11, 12, 14, 66, 68, 78

C

Cancer potency factor (CPF). *See* Risk
    assessment
Cancer slope factor (CSF). *See* Risk
    assessment
Cauchy distribution, 79, 212
CDF. *See* Cumulative distribution function
    (CDF)

CDF plot, 12–14, 63, 67–68, 78, 151, 176–178,
    181, 225, 228–231, 241
Censored data, 151
    Box-Cox transformations, 183–184
    classification of, 175
    empirical CDF plot, 176–178
        comparing to hypothesized
            distribution, 179–180
        comparing two groups, 180–181
    estimating parameters, 184–187
    estimating quantiles, 108, 197
    goodness-of-fit tests plotting results of, 3,
        150, 151, 175, 203–206
    graphs for, 176
    hypothesis tests, 203–208
        goodness-of-fit, 149–162, 203–206
    left censored, 175, 177, 182, 184, 191,
        198, 200, 203
    multiply censored, 3, 175, 176, 203, 207
    prediction intervals, 198–200
    Q-Q plots, 3, 149, 176, 181–183
    singly censored, 175, 183, 184, 200, 203,
        206, 207
    tolerance intervals, 200–203
    type I censored, 175, 184, 203
Census, 26
Chen's t-test, 163, 166–168
Chi distribution, 79
Chi-square distribution, 79
Chi-square goodness-of-fit test, 150, 159, 161
Companion scripts, 3, 5–7
Companion textbook, 4
Compliance monitoring, 52
Confidence intervals
    binomial proportion, 29, 34–38, 102
        sample size for, 34–38
    mean of lognormal distribution, 101, 189
    mean of normal distribution, 29–34, 187
        sample size for, 30–34
    quantile (nonparametric), 108
        sample size for, 108
    quantiles
        of lognormal distribution, 105–106

of normal distribution, 103–105
    relation to tolerance intervals, 125
sample size for, 28–40
Corrective action monitoring, 102, 104
Coverage, 44–48, 101, 141–146, 148
Cox, 75, 207
Cumulative distribution function (CDF), 75,
    87
    computing, 87, 92, 93, 222
    lognormal distribution, 68, 92–93
    plotting, 68, 87, 96

D

Data quality objectives (DQO), 27–28, 61
Datasets
    aldicarb, 103, 142
    arsenic, 41, 42, 45, 47, 116, 118, 131
    arsenic3.df, 41, 45
    benzene, 34, 70, 71, 102, 159, 161, 212,
        231
    chrysene, 105, 106, 118, 121, 123, 124,
        145
    copper, 38, 39, 146–147, 180, 181, 203,
        207
    Distribution.df, 79, 150
    EPA.94b.tccb.df, 5–9, 65, 66
    EPA.92c.arsenic3.df, 41, 45
    EPA.92c.benzene1.df, 34, 70, 102
    EPA.92c.copper2.df, 38
    EPA.02d.Ex.9.mg.per.L.vec, 166
    EPA.09.Ex.21.1.aldicarb.df, 103, 142
    EPA.09.Ex.18.1.arsenic.df, 116–117
    EPA.09.Ex.17.3.chrysene.df, 105, 145
    EPA.09.Ex.18.2.chrysene.df, 118
    EPA.09.Ex.17.4.copper.df, 147
    EPA.09.Ex.15.1.manganese.df, 184–185
    EPA.09.Ex.19.5.mercury.df, 137
    EPA.09.Ex.10.1.nickel.df, 154, 156,
        157
    EPA.09.Ex.21.6.nitrate.df, 109
    EPA.09.Ex.16.1.sulfate.df, 30
    EPA.09.Ex.19.1.sulfate.df, 134–136
    EPA.09.Ex.18.3.TCE.df, 126
    EPA.09.Ex.22.1.VC.df, 52
    EPA.09.Ex.18.4.xylene.df, 127
    Helsel.Cohn.88.silver.df, 177
    Millard.Deverel.88.df, 180
    Modified.TcCB.df, 181–182
    NIOSH.89.air.lead.vec, 101
    phosphorus, 170, 172
    silver, 177–180
    sulfate, 30, 32, 134–136

TcCB (censored), 8, 9, 65, 182, 184, 189,
    190, 194, 195, 197, 200,
    202–204
1,2,3,4-Tetrachlorobenzene (TcCB),
    7–10, 12–17, 19–23, 66–69,
    72, 73, 76, 77, 88, 89, 97–101,
    108, 146, 151–153, 159, 162,
    164, 165, 181–184, 189–191,
    200, 203, 206
Total.P.df, 147, 170–171
Data transformations. *See* Box-Cox
    transformations
Delta distribution. *See* Zero-modified
    lognormal distribution
Delta method, 189
Descriptive statistics. *See* Summary statistics
Design. *See* Sampling design
Distribution. *See* Probability distribution
Distribution parameters. *See* Parameters
Dose-response assessment. *See* Risk
    assessment
Dose-response curve. *See* Risk assessment
DQO. *See* Data quality objectives (DQO)

E

EDA. *See* Exploratory data analysis (EDA)
Empirical CDF plot, 12–14, 67–68, 78, 151,
    176–178, 181, 225, 228–231, 241
    censored data, 176–178, 181
    comparing to hypothesized distribution,
        63, 176, 179–180
    comparing two empirical CDFs, 61
    TcCB data, 12, 68, 151, 181
Empirical distribution, 213, 214, 230, 231
Empirical PDF plot, 63, 66–67, 78
Environmental statistics, 1–2, 4, 24, 79, 94,
    172
Environmental studies, 26–27, 36, 163, 169
    common mistakes, 26–27
EnvStats
    companion scripts, 5–7
    help, 5
    installing, 4
    intended users, 3
    masking, 6–7
    starting, 4–5
    system requirements, 4
    tutorial, 7–24
    unloading, 7
Estimating parameters
    binomial proportion, 102
    censored data, 184, 185, 187–189

lognormal distribution, 99–101, 184–187
    normal distribution, 97–99, 184–187
Estimating quantiles, 22–23, 103–106, 108,
        112, 197
    censored data, 197–198
    lognormal distribution, 105–106
        censored data, 188–190
    nonparametrically, 107–112
    normal distribution, 103–105
Exploratory data analysis (EDA), 2, 63
Exponential distribution, 79
Exposure assessment. *See* Risk assessment
Extreme value distribution, 2, 150
    goodness-of-fit test for, 159–161

F

F distribution, 92
Functions
    aovN, 50
    aovPower, 50
    attach, 4
    boxcox, 63
    boxcoxCensored, 176
    boxplot, 12
    cdfCompare, 63
    cdfCompareCensored, 176
    cdfPlot, 12, 63
    chenTTest, 163
    ciBinomHalfWidth, 31, 34, 36
    ciBinomN, 29, 34
    ciNormHalfWidth, 29, 31
    ciNormN, 29, 30
    ciNparConfLevel, 29, 38
    ciNparN, 29, 38
    ciTableMean, 29, 30
    ciTableProp, 29, 34
    dabb, 87, 88
    detach, 7
    dlnormAlt, 90
    dnorm, 87
    eabb, 97
    eabbCensored, 176
    ebinom, 102
    ecdfPlot, 12, 63
    ecdfPlotCensored, 188, 189
    egamma, 17, 18, 101
    elnorm, 98
    elnormAlt, 101
    elnormAltCensored, 188, 189
    elnormCensored, 184–187
    enorm, 97, 98
    epdfPlot, 63, 66, 88

    eqlnorm, 22, 106
    eqnorm, 103, 104
    eqnpar, 103, 110
    gofGroupTest, 150, 151, 154
    gofTest, 150, 151, 159
    gofTestCensored, 176, 203
    head, 8
    help, 5
    hist, 10
    kendallSeasonalTrendTest, 173
    kendallTrendTest, 170
    legend, 14, 33, 37, 43, 44, 47, 49, 60, 240
    library, 4
    linearTrendTestN, 50
    linearTrendTestPower, 50
    linearTrendTestScaledMds, 50
    oneSamplePermutationTest, 163
    pabb, 87, 88
    pdfPlot, 63, 66, 88
    plnormAlt, 88 (nf)
    plot, 14, 20, 32, 36, 48
    plotAovDesign, 50
    plotCiBinomDesign, 29, 34
    plotCiNormDesign, 29, 30
    plotCiNparDesign, 29, 38
    plotLinearTrendTestDesign, 50
    plotPredIntLnormAltTestPowerCurve, 51
    plotPredIntNormDesign, 40, 41
    plotPredIntNormSimultaneousTestPower
        Curve, 59, 60
    plotPredIntNormTestPowerCurve, 50
    plotPredIntNparDesign, 40, 43
    plotPropTestDesign, 50
    plotTolIntNormDesign, 45
    plotTolIntNparDesign, 45, 47
    plotTTestDesign, 49, 52
    plotTTestLnormAltDesign, 49
    pnorm, 87
    ppointsCensored, 176
    predIntGamma, 114
    predIntGammaAlt, 114
    predIntGammaSimultaneous, 114
    predIntGammaSimultaneousAlt, 114
    predIntLnorm, 114, 120, 121
    predIntLnormAltSimultaneousTestPower,
        51
    predIntLnormAltTestPower, 51
    predIntLnormSimultaneous, 114, 134
    predIntNaprConfLevel, 40
    predIntNorm, 40, 41, 114, 121
    predIntNormHalfWidth, 40
    predIntNormK, 114
    predIntNormN, 40, 41

predIntNormSimultaneous, 50, 58, 114
predIntNormSimultaneousK, 114
predIntNormSimultaneousTestPower, 50, 58
predIntNormTestPower, 50
predIntNpar, 114
predIntNparConfLevel, 40, 43
predIntNparN, 40, 43
predIntNparSimultaneous, 40
predIntNparSimultaneousConfLevel, 40
predIntNparSimultaneousN, 40
predIntPois, 114
propTestMdd, 50
propTestN, 50
propTestPower, 50
qabb, 87, 88
qlnormAlt, 93
qnorm, 87
qqPlot, 63
qqPlotCensored, 176, 199
qqPlotGestalt, 63
quantileTest, 163
rabb, 87, 88, 216
rlnormAlt, 94
rnorm, 87, 216
runif, 216
search, 4
serialCorrelationTest, 169, 173
set.seed, 94, 216
signTest, 163
simulateMvMatrix, 88, 95, 222
simulateVector, 88, 221
stripChart, 10, 63, 66, 149, 163, 165
summary, 7, 63, 65, 163, 164, 214, 224, 230, 234
summaryFull, 9, 63
summaryStats, 9, 63, 149, 163, 164
tolIntGamma, 141, 146
tolIntGammaAlt, 141, 146
tolIntLnorm, 141
tolIntLnormAlt, 141
tolIntNorm, 45, 141
tolIntNormHalfWidth, 45
tolIntNormK, 141
tolIntNormN, 45
tolIntNpar, 45, 47, 141
tolIntNparConfLevel, 47
tolIntNparCoverage, 45, 47
tolIntNparN, 45, 47
tolIntPois, 141
tTestAlpha, 49, 52
tTestLnormAltN, 49
tTestLnormAltPower, 49

tTestLnormAltRatioOfMeans, 49
tTestN, 49, 52
tTestPower, 49, 52
tTestScaledMdd, 49, 52
twoSampleLinearRankTest, 163
twoSampleLinearRankTestCensored, 207
twoSamplePermuationTestLocation, 163
twoSamplePermuationTestProportion, 163
varGroupTest, 163
varTest, 163
wilcox.test, 24

G

Gamma distribution, 3, 14, 17, 19, 21, 22, 90–91, 101–102, 106–107, 114, 122–124, 136, 141, 146, 190–191
Generalized extreme value distribution, 80
Generalized pivotal quantity, 187, 194, 195, 200
Geometric distribution, 80
Goodness-of-fit tests, 3, 20–22, 124, 136, 149–162, 166, 172, 175, 176, 180, 183, 203–208, 229
    censored data, 150, 175, 176, 183, 203
        plotting results of, 181
    chi-square, 150, 159, 161
    continuous distribution, 3, 150, 159
    for group normality, 150, 154, 158
    Kolmogorov-Smirnov, 150, 159, 161, 162
        to compare two samples, 150
    for normality, 150–154, 176, 180
    plotting results of, 20, 153, 158, 162, 206
    probability plot correlation coefficient, 73, 150, 151, 159
    Shapiro-Francia, 149, 150, 180, 203
    Shapiro-Wilk, 19–21, 76, 150, 152, 158, 166, 183, 203
Groundwater protection standard (GWPS), 52, 53, 102, 103, 142
GWPS. See Groundwater protection standard (GWPS)

H

Hazard identification. See Risk assessment
Help system. See EnvStats
Histogram, 10, 11, 14, 64, 66–68, 75, 88, 89, 98–100, 169, 176, 220, 225, 228–230, 241
Hot-measurement comparison, 102, 144, 147

Hypergeometric distribution, 80
Hypothesis tests, 2, 3, 28, 49–61, 66, 75, 78,
        149–173, 203–208
    Bartlett's test, 163
    censored data, 203
    Chen's t-test, 167
    comparing groups, 229
    goodness-of-fit, 124, 149, 154, 166,
        203–206
        censored data, 203
        group normality, 154
        normality, 154
    Levene's test, 163
    linear rank tests, 207
        censored data, 207
    permutation test, 3
    quantile test, 3, 207
    sample size and power, 3, 49–51
    serial correlation, 169
    sign test, 163
    Student's t-test, 49, 52, 163–166
        sample size and power, 49, 52
    trend, 3, 169–172
    variance test, 163

I

ILCR. *See* Incremental lifetime cancer risk
        (ILCR)
Important parameters, 226
Imputation, 185, 186, 188, 201
Incremental lifetime cancer risk (ILCR), 212,
        226, 232
Installing. *See* EnvStats
Inverse transformation method, 216, 217,
        220

J

Judgment sampling. *See* Sampling

K

Kaplan-Meier, 176, 183, 191, 194
Kendall test for trend, 149, 169, 171, 172
    seasonal, 169, 171, 172
Kolmogorov-Smirnov test, 150, 159, 161,
        162
    to compare two samples, 150

L

Land's method, 99, 101

Latin Hypercube sampling, 217–223, 228,
        230, 237, 239, 241
    properties of, 222
    replicated, 222, 230
    vs. simple random sampling, 220–221
Law of Large Numbers, 214
Left censored, 175, 177, 182, 184, 191, 198,
        200, 203
Linear rank tests, 23, 24, 163, 166–169, 176,
        207
    censored data, 207
Logistic distribution, 80
Lognormal distribution, 2, 3, 14–16, 19, 20,
        22, 49, 51, 64, 68–70, 75, 80,
        88–95, 97, 99–102, 105–106, 113,
        114, 118–122, 124, 133–135, 141,
        144–146, 149–152, 157, 159, 166,
        176, 179, 180, 182, 184–190, 192,
        194–197, 199, 200, 203, 223, 230,
        231, 236, 237, 239
    confidence interval for mean, 101, 102,
        176, 184, 189, 191, 192, 194
    cumulative distribution function, 13, 14,
        92–93, 179, 180
    estimating parameters for, 99–101,
        184–188
        censored data, 184, 188
    estimating quantiles for, 104–105
        censored data, 197
    prediction intervals, 51, 119–122,
        133–135, 144
    probability density function, 88–90
    simultaneous prediction intervals, 51,
        114, 133–135
    three-parameter, 80, 150
    tolerance intervals, 141
    truncated, 80, 236, 237, 239
Lognormal mixture distribution, 80

M

Mann-Kendall test for trend, 169
Masking, 6–7
MASS package, 94, 214, 222
Maximum concentration limit (MCL),
        102–104, 142, 144
MCL. *See* Maximum concentration limit
        (MCL)
Mean
    confidence interval for sample size for, 9,
        10, 16–18, 28–34, 63, 66, 73,
        97, 101, 102, 163, 165, 184,
        187–189, 191, 192

lognormal distribution, 16, 95, 101, 184, 223
normal distribution, 29–34, 51–55, 70, 95, 114
  test on, 51–55
Monte Carlo simulation, 2, 197, 211–242
  input parameters, 212, 228
  Latin Hypercube sampling, 217–222
  origin of term, 231
  sensitivity analysis, 224–231
    methods for, 227–229
  sum of two normal random variables, 213–215
  uncertainty analysis, 224–231, 234
    methods for, 230–231
Multivariate normal distribution, 94
  generating random numbers from, 94
Multivariate random numbers, 94–95

N

Negative binomial distribution, 80
Nitrate, 108, 109, 198
Non-central beta distribution, 79
Non-central chi-square distribution, 79
Non-central F distribution, 79
Non-central Student's t distribution, 89
Nonparametric
  confidence interval for quantile, 28, 38–40, 103, 108
  estimate of quantiles, 107–112, 193, 198
  prediction intervals, 125–128, 146, 198, 200
    sample size for, 125
  simultaneous prediction intervals, 136–14
  test to compare two groups, 207–208
    censored data, 207–208
    linear rank test, 207
    quantile test, 207
  tolerance intervals, 47–49, 51, 146–148, 200, 203
    sample size for, 48
Normal distribution, 29, 30, 40–43
  confidence interval for mean, 96
    sample size for, 32
  estimating parameters for, 96
  estimating quantiles for, 103–105
  goodness-of-fit tests, 151, 203
    group, 151
  prediction intervals, 40–43, 50, 116–119
  sample size for, 40, 44, 49
  simultaneous prediction intervals, 57

sum of two normal random variables, 213–215
test on mean, 51–55
  sample size and power, 51–52
tolerance intervals, 45–47, 142–144
  sample size for, 45
truncated, 80
Normal mixture distribution, 80

P

Parameters
  confidence intervals for, 97, 98, 100–104, 106, 108, 112
  estimating, 16–19
    censored data, 184–193
  important, 102
  sensitive, 225–226
Pareto distribution, 60
PDF. *See* Probability density function
Percentiles. *See* Quantiles
Period, 25, 26, 52, 53, 79, 170, 216, 233
  of random number generator, 216, 223
Permutation tests, 3, 163, 166
Plotting
  boxplot, 66
  censored data, 176–178
  cumulative distribution function, 63
  empirical CDF, 67–68
    censored data, 180
  goodness-of-fit test results, 149–154
    censored data, 151
  histogram, 98, 100
  PPCC *vs.* Box-Cox transform, 77, 183
  probability density function, 89, 90
  Q-Q plot
    building gestalt for, 63, 73–74
    comparing groups, 63, 72, 73
  sample size and power for hypothesis tests, 52
  sample size for confidence intervals, 28–30, 32, 34, 36, 38, 40
  sample size for prediction intervals, 40–44
  sample size for tolerance intervals, 44–49
  Tukey mean-difference Q-Q plot, 68, 73, 77, 151, 154
Plotting positions, 68
  censored data, 176, 177, 181
Poisson distribution, 70, 75, 114, 141, 159–161
  Q-Q plot for, 70, 72

Population
    definition of, 1, 25
Power, 3, 18, 19, 27, 28, 49–61, 75–77, 113,
        118, 120, 122–124, 129, 132–134,
        140, 148, 183, 184, 231
PPCC. *See* Probability plot correlation
        coefficient (PPCC) test
Prediction intervals, 3, 28, 40–44, 57–61,
        113–140, 142, 148, 198–200
    censored data, 165
    definition of, 109
    for future means, 116, 118–119
    for future median, 127, 136
    gamma distribution, 114, 122–124
    lognormal distribution, 119–122
    nonparametric, 125–128
        sample size for, 128
    normal distribution, 116–119
        sample size for, 117–119
    sample size for, 40–44
    simultaneous, 117
        lognormal distribution, 133–135
        nonparametric, 136–140
        normal distribution, 131–133
        rules for, 135, 136
Probability
    definition of, 1
Probability density function
    computing, 2
    lognormal distribution, 88–90
Probability distribution
    abbreviations for, 79, 80, 87, 88, 96
    parameters, 79, 80, 91, 94, 96
Probability plot. *See* Q-Q plot
Probability plot correlation coefficient
        (PPCC) test, 76, 77, 150, 151, 159,
        183, 184
    group, 150
Probability sampling. *See* Sampling
Profile likelihood, 187–191
Pseudo-random numbers. *See* Random
        numbers

Q

Q-Q plot, 14, 15, 68, 77, 78, 151, 154
    arbitrary distribution, 14–16, 69
    building gestalt for, 73–74
    censored data, 176, 181–183, 186
    comparing two groups, 63
    estimating mean and standard deviation
        from, 30, 31
    lognormal distribution, 69–70

    normal distribution, 69–70
    Poisson distribution, 70, 75, 159
    TcCB data, 14–16, 69
    Tukey mean-difference, 74, 77
        building gestalt for, 73–74
Quantile plot. *See* Empirical CDF plot
Quantile-Quantile plot. *See* Q-Q plot
Quantiles
    computing, 83, 93, 182
    confidence interval for, 98, 100–104,
        106–108, 110–112
        relation to tolerance intervals, 106
    definition of, 97
    in environmental standards, 102
    estimating, 22, 103–106, 108, 112
        censored data, 22, 103–106, 108,
            112, 197
    estimating nonparametrically, 103,
        107–109
    lognormal distribution, 93
    nonparametric confidence interval for,
        108
        sample size for, 108
    normal distribution, 97–99
    relation to percentiles, 93, 102
Quantile test, 3, 23, 149, 163, 168–169, 207

R

Random numbers, 2, 87, 88, 94–96, 212–214,
        216–224
    algorithm for generating, 136
    inverse transformation method, 216, 217
    Latin Hypercube sampling, 117–122
    lognormal distribution, 94, 95
    in Monte Carlo simulation, 222
    multivariate based on rank correlation,
        95–96
    multivariate normal, 94–95
    period of generator, 216
    pseudo-random numbers, 94, 216
    seed for, 197
    uniform distribution, 216
Random sampling. *See* Sampling
Rank correlation, 88, 95, 222–224, 241
    generating multivariate random numbers
        based on, 95–96
Regression on order statistics (ROS), 183,
        185, 186, 188, 196, 202
Risk, 108–109, 111
Risk assessment, 1, 3, 94, 211–213, 231–241
    building model for, 235–236
    cancer potency factor, 233, 236

cancer risk for benzene in soil, 232
cancer risk from benzene in soil, 212
cancer slope factor, 237
definition of, 232
definition of risk, 232
detailed example, 232
dose-response assessment, 195, 233, 234
dose-response curve, 233, 234
exposure assessment, 234
guidelines for, 130
hazard identification, 233, 234
incremental lifetime cancer risk, 212
model for, 195, 197, 211, 212, 233, 234
PCE in groundwater
point estimates of risk, 212, 226, 237
probabilistic risk assessment, 212, 232
problems with, 233, 235
risk characterization, 234, 235
risk communication, 235
risk management, 234–235
sensitivity analysis, 231, 234, 241–242
three major steps of, 233
uncertainty analysis, 230–232
Risk characterization, 234, 235
Risk communication, 235
Risk management, 234
ROS. *See* Regression on order statistics

S

Sample
definition of, 1, 26
physical, 26
Sample size
confidence intervals, 28–40
hypothesis tests, 49–61
prediction intervals, 40–44
tolerance intervals, 44–48
Sampling
judgment, 26
probability, 26, 61
random, 26, 61, 216, 220–221, 230
Sampling design, 2, 3, 25–27, 29, 40, 45, 49, 50
Seasonal Kendall test for trend, 169, 171, 172
Seed, 55, 56, 72, 73, 94, 95, 195–197, 202, 214, 216, 221, 223, 239, 240
for random number generation, 216
Sensitive parameters, 225–226
Sensitivity analysis
important *vs.* sensitive parameter, 225
methods for, 227–229
Serial correlation test, 169, 173

Shapiro-Francia test, 149, 150
group, 150
Shapiro-Wilk test, 19, 151, 152
group, 151
Sign test, 163
Simultaneous prediction intervals
lognormal distribution, 114, 133–135
nonparametric, 136–140
normal distribution, 131–133
rules for, 135, 136
Singly censored, 175, 183, 184, 200, 203–207
Starting EnvStats. *See* EnvStats
Statistics
definition of, 1
Student's t distribution, 80
Student's t-test, 49, 51, 52, 121, 163–166
sample size and power, 49
Summary statistics, 2, 7, 9, 10, 63–66, 78, 163, 164, 214, 224, 225, 230, 241
TcCB data, 65–66
System requirements. *See* EnvStats

T

Table of contents, 8, 28–30, 40, 43–45, 47–49, 51, 61, 63–65, 78, 79, 87–88, 96, 97, 103, 112–114, 125–127, 136, 137, 141, 146–148, 150, 159, 163, 170, 172, 173, 176, 177, 180–185, 187–190, 194, 197, 198, 200, 203, 207, 209, 214, 223, 236, 237
Technical support, 1
Three-parameter lognormal distribution, 150
Tolerance intervals
β-content, 142–145, 148
β-expectation, 142
censored data, 175, 176, 200–203
coverage, 141–146, 148
definition of, 141
lognormal distribution, 114, 118–122, 124, 133–136, 141, 144–146
nonparametric, 114, 125–128, 130, 136–141, 146–148
sample size for, 47, 125, 126, 128, 147
normal distribution, 114, 116–120, 122, 133, 140–144
sample size for, 45, 117–119, 143
relation to confidence intervals for, 113
quantiles, 114
sample size for, 119, 122, 124
Transformations. *See* Box-Cox transformations

Trend, 3
    tests for, 50, 149, 169–173, 211
Triangular distribution, 236
Truncated lognormal distribution, 236, 237,
    239
Truncated normal distribution, 80
Tukey mean-difference plot. *See* Q-Q plot
Tutorial, 1, 7–24
Type I censored data, 184, 186, 187, 189,
    190, 192–194, 196, 197, 199,
    201–205

U

Uncertainty, 26, 27, 61, 142, 212, 213, 234,
    236–242
Uncertainty analysis
    definition, 225
    methods for, 19, 230–232
    uncertainty *vs.* variability, 188, 226–227
Uniform distribution, 154, 216, 231, 237
    random numbers from, 216
Unloading EnvStats. *See* EnvStats

V

Variability, 27–30, 64, 79, 128, 134, 137,
    154, 211, 225–227, 230, 232,
    236–242
Variance test, 163
version, 4–6, 198

W

Weibull distribution, 80
Wilcoxon rank sum distribution, 23, 24, 80,
    168, 207, 229
Wilcoxon rank sum test, 23, 24, 121, 168,
    169, 207, 229

Z

Zero-modified lognormal distribution, 2, 80,
    150
Zero-modified normal distribution, 80, 150

Printed by Books on Demand, Germany